Science and Humanity

The Extraordinary Life of Joseph H Hamilton

We have had a significant and wonderful scientific and personal time together. — I hope coexistence in the Hg isotope to be but recoil mass spectrometer ever built.

On a personal note, I still treasure the beautiful 1851 pistol you got for me. I deeply appreciate the remark of your "Father-in-law" when we met who said "I ery day 3 thing important in his life God, Family and Joe Hamilton and I would not want to rank them"

Thank you for your special article for this book.

Joe 6/23/2025

Science and Humanity

The Extraordinary Life of Joseph H Hamilton

editors

Da Hsuan Feng

Drexel University, USA

Mark A Riley

Florida State University, USA

Lee L Riedinger

University of Tennessee Knoxville, USA

William H Brantley

Furman University, USA

World Scientific

NEW JERSEY • LONDON • SINGAPORE • GENEVA • BEIJING • SHANGHAI • TAIPEI • CHENNAI

Published by

World Scientific Publishing Co. Pte. Ltd.

5 Toh Tuck Link, Singapore 596224

USA office: 27 Warren Street, Suite 401-402, Hackensack, NJ 07601

UK office: 57 Shelton Street, Covent Garden, London WC2H 9HE

Library of Congress Control Number: 2024949498

British Library Cataloguing-in-Publication Data
A catalogue record for this book is available from the British Library.

SCIENCE AND HUMANITY
The Extraordinary Life of Joseph H Hamilton

ISBN 978-981-12-9695-6 (hardcover)
ISBN 978-981-12-9696-3 (ebook for institutions)
ISBN 978-981-12-9697-0 (ebook for individuals)

For any available supplementary material, please visit
https://www.worldscientific.com/worldscibooks/10.1142/13953#t=suppl

Desk Editor: Muhammad Ihsan Putra

Typeset by Stallion Press
Email: enquiries@stallionpress.com

Preface

The career of Joseph Hamilton is multi-dimensional, prodigious in accomplishment, and unique. To review the career of such a world-recognized scientist, one must go beyond the usual summary of scientific output and look far more broadly at what he has meant to science, its institutions, and his students, colleagues, and family. This is what we have attempted to do in this volume.

In anticipation of Joe's retirement from 64 years of distinguished service at Vanderbilt University, three of us (Da Hsuan Feng, Lee Riedinger, and Mark Riley) decided to organize an event in which we and many others could recognize a person that we associate with the descriptors *great*, *extraordinary*, and *legend*. On June 8, 2022, we held a global ZOOM forum entitled *Science and Humanity: The Extraordinary Life of Joseph H. Hamilton* to honor him. This special event was recorded and is available at: https://gradschool.fsu.edu/science-and-humanity-extraordinary-life-joseph-h-hamilton.

In this historic gathering, we were touched by so many who spoke about Joe's life as a truly incredible human being and how he has always had a deep concern, even love, not only for his colleagues (including hundreds of students) but also for humanity in general. From all the comments made by the members of the scientific and non-scientific communities, including heartfelt ones from Joe's family, we understood the huge impact he has had on people around the globe, those in science, and those far beyond.

After the forum, we realized the importance of recording this outpouring of stories and sentiment in a written format, so that many more can understand the breadth and impact of Joe's career. It is important for future generations of scientists to learn about his career and to grasp what it really means to be a scientist and a leader.

This book is much more than a *Festschrift* recounting his scientific output since it celebrates many more facets of Joe's long career. We start with a technical chapter on Joe's major scientific discoveries and his development of new facilities, followed by chapters devoted to his teaching, his mentoring of graduate students and postdoctoral fellows, his many international collaborations, his political and fundraising activities, his and his wife's science and religious writings and lectures, remembrances from family members including personal reflections by Joe himself, poems written about Joe, plus Joe's own brief summary of the present endeavor. In the appendices, a URL link to a full listing of his curriculum vitae is presented, including details of more than 1,200 publications, 450 invited talks, plus 100 graduate students and 120 postdoctoral scholars mentored. Also included is the State of Tennessee certificate honoring Joe for his accomplishments. Finally, some photographs of Joe through his illustrious career are included.

Joining the organizers of the June 2022 forum was one of Joe's early Ph.D. students, William Brantley of Furman University, to assist in the construction of the book and the arduous editing process. We would also like to acknowledge the contributions of Ms. Peggy Lucas-McGowan to the event poster and cover design of this volume. We are very grateful for the cooperation of one of the world's most distinguished scientific publishing houses and the support of Dr. K.K. Phua of the World Scientific Publishing Corporation in making this project a reality. The editorial assistance of Soh Yong Qi has been invaluable. We appreciate the help of the many people who contributed to this volume in honor of Joseph Hamilton.

Da Hsuan Feng
Mark Riley
Lee Riedinger
William Brantley

Contents

Joseph H. Hamilton and Vanderbilt University

John Geer

*Ginny and Conner Searcy Dean of the College of Arts and Sciences,
Vanderbilt University, Nashville, TN 37235, USA
john.g.geer@vanderbilt.edu*

*Dean of the College of Arts and Sciences at Vanderbilt University,
Dr. John Geer's introductory remarks to the global Zoom celebratory
event on the extraordinary life and career of Joseph H. Hamilton.*

1. Introduction

Hello, I'm John Geer, and I am the Ginny and Conner Searcy Dean of the College of Arts and Sciences at Vanderbilt University. I have the honor, and distinct privilege, of being both a colleague and a friend of Joe Hamilton. When you think about Joe's career, one of the things you might immediately be struck by is how long of a career he has enjoyed. It's been amazing. Consider that he has been at Vanderbilt during six of our nine Chancellors. But the length of his career is not what really matters. It's the impact of his work, whether it's with students, his research, his outreach globally and across the country, or the kind of humanitarian efforts he undertook. He has left a mark that few have. The best part about it is while he is moving to emeritus status, he is appointed as Distinguished Research Professor, so his career is not over! He can still do his research, still interact with students, and still be part of our community, and for that, I am and will be very grateful. So, it is a pleasure to just say a few words and express my deep admiration for Joe Hamilton. He is a legend here at Vanderbilt, and he will continue to be a legend, and we are certainly lucky to count him as a friend and a colleague. I wish the very best for today's events. You be well Joe and hope to see you soon.

Thank you.

Educator and Researcher Extraordinaire

Lee L. Riedinger

*Physics Department, University of Tennessee,
Knoxville, TN 37996-1200, USA
lrieding@utk.edu*

The research career of Joseph Hamilton has evolved over 64 years in five major phases. The research output has been prolific and his leadership in forming new consortia and building new instruments impressive and unique. Education of students has been an outstanding constant in his career. Many major awards have recognized his impact on the field of nuclear physics.

1. Introduction

Legends in various fields are often described by the 'numbers' they achieve throughout their careers and by the impact they have had on people, ideas, and directions in their areas of expertise. In each of these ways, Professor Joseph Hamilton is a legend in his field of nuclear physics and in his impact on students, colleagues, and organizations. As one of his early doctoral students and as a close colleague for decades, I can attest to his status as a legend.

2. Background

The birthplace of Joe Hamilton was Ferriday, Louisiana, the same town that produced another legend, Jerry Lee Lewis, the rock and roll phenom. Joe's long career took a far different pathway than Lewis', as he went to school at Mississippi College (bachelor's degree in 1954) and Indiana University for a doctorate in physics in 1958. Soon thereafter he joined the faculty of Vanderbilt University, albeit with his first faculty year spent at the Siegbahn Institute in Stockholm. His service on the Vanderbilt faculty extended 64 years until his retirement in 2022. When queried about his longevity in physics, Joe often explains that his father worked as an active Baptist minister well past age 90.

Those 64 years have been impactful and productive almost beyond reason, resulting in more than 1,200 publications (and still counting), around 100 graduate students educated with Vanderbilt degrees, 120 post-doctoral research associates mentored, and 450 invited talks presented. In addition to these many accomplishments related to nuclear physics, Joe has been an undergraduate teacher of high repute at Vanderbilt. For decades he taught a physics course for non-science majors. His passion for physics and his knack for showmanship in demonstrations made him an extremely popular teacher in a discipline (physics) not known for attracting the masses. In the process, he taught perhaps 10,000 non-science students, a very impressive statistic. He has received awards for his innovative teaching.

Not surprisingly, a career so steeped in accomplishment has been recognized and rewarded in various ways. Joe has received eight honorary degrees, including five doctorates from institutions abroad: Johann Wolfgang Goethe Universitat, Frankfurt, Germany (1992); University of Bucharest, Romania (1999); St. Petersburg State University, Russia (2001); Joint Institute for Nuclear Research, Russia (2004); Shukla University, Raipur, India (2006). In addition, there has been an impressive array of honors and awards, too numerous to recount. Included in this list are many Vanderbilt awards for teaching, research, and service; an American Association for the Advancement of Science Award for International Cooperation (1996); the Ilkovic Gold Medal for Career Achievements in Science, Slovak Academy of Sciences (2001); an

International Scientific and Technological Cooperation Award of the People's Republic of China (2002); the G. N. Flerov Prize for major achievements in Nuclear Physics Research, given by the Joint Institute for Nuclear Research, Russia (2003); and awards for research, service, and teaching by the Southeastern Section of the American Physical Society. These awards attest to decades of outreach to institutions in China, India, and Eastern Europe.

3. Research

A research career covering 64 years has not been static in direction but has evolved over five main emphases in nuclear physics. Along with Joe's evolving research interests has been his frequent leadership in organizing consortia and building instruments that would facilitate his new research foci. In the process, many physicists and institutions have greatly benefited from his activism and leadership. Joe's many graduate students and postdocs have received training at the forefront of nuclear physics by their participation in these different research emphases. Many of these early-career associates of Joe Hamilton have proceeded to build long careers of their own as a result of his mentorship, and quite a few have ended up retiring before he finally did in 2022. Joe's longevity has been amazing.

3.1. *Radioactive decay*

The earliest emphasis of Joe Hamilton's research career was the study of nuclear states through radioactive decay. This direction produced around 133 published papers, the first in 1956 from his work as a grad student at Indiana.[1] This work, as for many of his early papers, involved the use of magnetic beta-ray spectrometers, first at Indiana and then at Vanderbilt using one he designed. One of the co-authors in Ref. 1 is Russell Robinson, then a fellow grad student, later a researcher at Oak Ridge National Laboratory (ORNL) and one who brought Joe into phase two of his research career.

In 1966, I was the first of many Vanderbilt graduate students that Joe sent to ORNL to work on dissertation research in his first major collaboration with staff scientists there. My dissertation involved the study of gamma-rays emitted in the decay of ^{152}Eu and ^{154}Eu, both long-lived radioactivities. A new technical aspect of this work was the development at ORNL of one of the first high resolution gamma-ray detectors: Ge(Li) — lithium drifted germanium crystals. Though small in volume (6 cm^3), the high energy resolution permitted the detection of many new gamma-ray transitions that were earlier unseen due to the previous generation of much lower resolution NaI scintillator counters. These experiments resulted in the second[2] of Joe's 25 *Physical Review Letters* throughout his long career.

One purpose of this work was to study the β- and γ-vibrational bands in ^{152}Sm and ^{154}Gd, two nuclei ($N = 90$) at the beginning of the rare-earth region of prolate deformation.[3] The branching ratios deviated from the predictions of the Bohr-Mottelson collective model but could be better explained by allowing the wave functions of these two excited vibrational bands to mix with the ground-state band and with each other. This work set the pattern for many subsequent studies of vibrational bands in other deformed nuclei involving the Vanderbilt group and others, including the first angular correlation studies with Ge(Li) detectors to determine the multipolarities of the transitions. One such crucial experiment they did found that the transition from the 2^+ member of the β band to the 2^+ of the ground band in ^{154}Gd was in fact 99% E2.[4] This was an important and timely experiment, since Ben Mottelson surmised (in his talk at the 1967 Tokyo conference) that this transition would be 50% M1 in nature, in order to rectify the branching ratios with his collective model. Much theoretical work on the nature of β bands in deformed nuclei has resulted from this work by the Hamilton group.

3.2. *Coulomb excitation work at ORNL*

The work of the Hamilton group in studying the properties of vibrational bands in deformed nuclei by radioactive decay led to the second phase of Joe's long research career — Coulomb excitation to measure the collectivity of these excited rotational bands. To carry out this series of measurements, Joe worked closely with his former graduate-student friend,

Russell Robinson, by then a staff researcher at ORNL. This series of experiments utilized beams from the EN tandem accelerator and involved a series of graduate students from the Hamilton group. The first joint publication[5] used accelerated alpha particles to excite levels in ^{178}Hf where three known 2^+ states in the 1.2 to 1.6 MeV range were candidates to be members of a collective $K = 0$ β-vibrational band, perhaps akin to those seen at lower excitation energies in $N = 90$ Sm and Gd nuclei (see Fig. 1). However, these Coulomb excitation measurements showed that none of the candidates have an enhanced B(E2) value (indicative of collectivity) and thus could not be the hoped-for β-vibrational band. This experiment demonstrated for the first time that the collectivity that is concentrated in the low-lying 0^+ band (around 680 keV) at $N = 90$ deformed nuclei must be spread over many higher lying 0^+ bands in the middle ($N = 106$) of the rare-earth deformed region, therefore no β band.

This series of experiments at ORNL produced another important result led by a Hamilton graduate student.[6] Their Coulomb excitation measurements on ^{180}Hf led to the first evidence for a *negative* hexadecapole moment in a deformed nucleus. Hafnium nuclei (like many) have sizeable positive (football shaped) quadrupole deformations (β_2) but also much smaller β_4 values, generally positive (a small bulge around the waist of the football) but surprisingly negative for ^{180}Hf indicating a peanut shape.

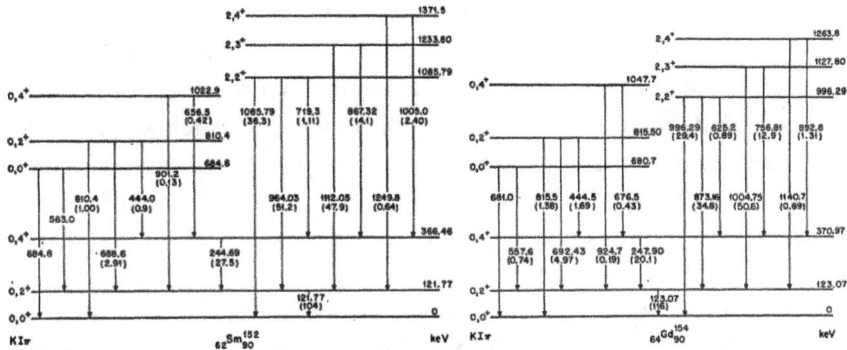

Fig. 1. Partial level scheme of ^{152}Sm and ^{154}Gd showing the transitions from the β ($K = 0$) and γ ($K = 2$) vibrational bands.

Other important experiments were performed by the Hamilton group at the ORNL EN tandem lab, e.g., landmark measurements on shape coexistence, as discussed below.

3.3. *UNISOR era*

Joe Hamilton's career has been defined not only by forefront research in nuclear physics but also by his leadership in forming new collaborations and building new instruments. A prime example of is the University Isotope Separator at Oak Ridge (UNISOR). In the decade of the 1960s he had led many studies of nuclei populated by radioactive decay and had developed collaborations at ORNL. In 1969 he led the first meeting at Vanderbilt discussing his idea to develop an on-line device to study radioactive nuclei with short half-lives and produced by heavy-ion induced reactions using beams from the Oak Ridge Isochronous Cyclotron (ORIC). A year later he organized a planning session at ORNL involving representatives from 10 universities, including Vanderbilt and the University of Tennessee (UT). This produced the plan to purchase and install on an ORIC beamline a mass separator that would allow the production and separation of short-lived radioisotopes — UNISOR. The purchase of this device was made possible by contributions from these 10 universities plus the State of Tennessee. UNISOR was the first *user facility* at ORNL and likely at any national laboratory of the Department of Energy — a facility purchased and operated by a set of universities. Now every national lab hosts a number of user facilities, and Joe Hamilton's ideas and leadership resulted in the first one.

The UNISOR collaboration led to 43 years of research at ORNL by the member universities and provided in some cases the first opportunities for faculty to engage in forefront funded research. Hundreds of experiments and publications resulted. The existing paradigm in 1970 was that every nucleus had one fixed shape as described by two distinguished theorists in Scientific American. No theorist had predicted that complex nuclei could have coexisting shapes. The Hamilton group led some of the earliest and most impactful experiments on the discovery of new isotopes ^{184,186,188}Tl and their decay to ^{184,186,188}Hg.[7] These experiments established energy levels built on two very different shapes — the near-spherical shape expected for Hg nuclei ($Z = 80$, two protons short of a closed shell) and an *intruding* band built on a deformed shape. This *coexistence* of shape features so close to a closed shell of protons was very surprising to many and a shock to some. The levels in the coexisting structures are shown in Fig. 2 and represent a classic example of a phenomenon of shape coexistence that has now been identified in many nuclei in regions off the line of stability. This was a landmark experiment, one of many in Hamilton's career. He had changed the paradigm of each nucleus having only one fixed shape to shape coexistence being possible in nuclei.

Around the same time, Hamilton's group was collaborating with Russell Robinson and colleagues at the ORNL EN tandem accelerator on equally surprising shape coexistence measurements in light Sr and Kr nuclei. They first studied ^{72}Se by the (^{16}O,2p) reaction and identified a rotational band based on a 0^+ state located at 937 keV and extending to spin 12.[8]

Next, they performed the first measurement of nuclear lifetimes of excited nuclear states populated in-beam. Using from a line-shape analysis of the Doppler-broadened γ-ray lines, they measured lifetimes in this band of energy levels indicative of a strongly deformed rotational band

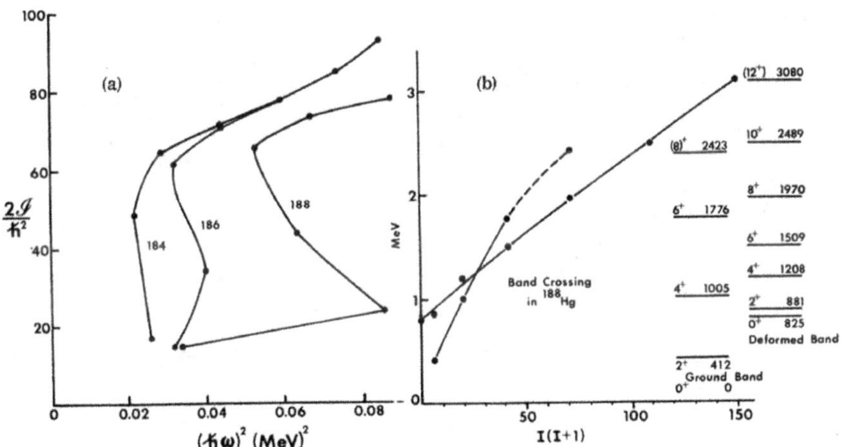

Fig. 2.   (a) Moment of inertia versus frequency squared for the yrast sequences of three Hg isotopes. (b) Level energy versus level spin squared for the crossing spherical and deformed structures in ^{188}Hg.

coexisting with the near spherical ground-state structure.[9] An even more surprising result came from their in-beam spectroscopy measurements of levels in ^{74,76}Kr at the ORNL and Köln EN tandem accelerators and ^{76}Rb decay experiments at UNISOR.[10] Their work demonstrated that the *ground states* of these light Kr isotopes are strongly deformed, while the expected near-spherical levels are elevated in energy. They explained the surprising appearance of these ground state deformed shapes as the result of energy gaps in proton and neutron single-particle diagrams at the same deformation — *reinforcing shell gaps*. As seen in Fig. 3, a gap appears in the Nilsson diagram of single-particle energies at a large prolate deformation when around 38 particles have been placed in these orbits. In fact, ^{74}Kr has 36 protons and 38 neutrons, thus achieving a prolate ground state in part due to the deformation driving nature of the $g_{9/2}$ orbital.

Hamilton's ideas on shape coexistence were initially controversial. At the 1976 *International Conference on Selected Topics in Nuclear Structure*

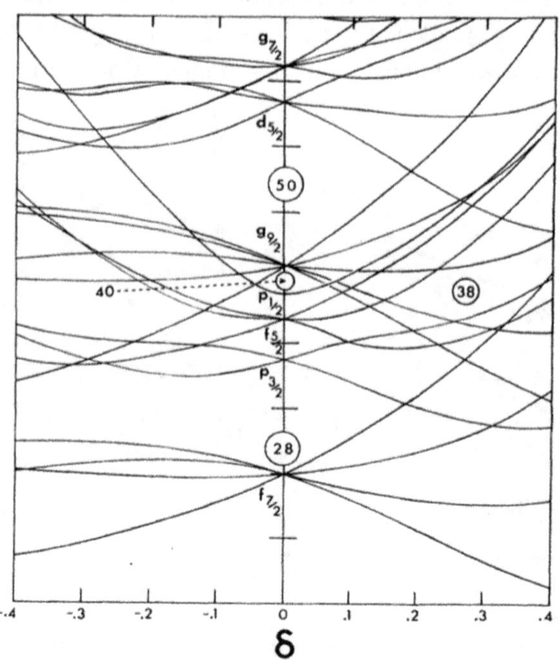

Fig. 3.   Nilsson diagram of single-particle energy levels versus deformation.

in Dubna, his invited talk was followed by a 30-minute discussion and debate about these new ideas, a highlight of that conference. By the time of the August 2018 Nuclear Structure conference at Michigan State University, this initially debated idea became commonplace, as many talks highlighted nuclear shape coexistence found in many different areas of the nuclear chart. Joe was right! In fact, V.G. Soloviev, the distinguished Russian theorist and chair of the Dubna conference, later told Hamilton that he had nominated Joe for a Nobel Prize in physics for the discovery of shape coexistence in complex nuclei, as also did a noted German theorist. Unfortunately, this special honor was not awarded.

3.4. *Spontaneous fission products*

An amazingly productive aspect of Joe Hamilton's career then focused on studying the structure of nuclei produced as fission products from the spontaneous fission of ^{252}Cf. Of course, a host of nuclei are produced in fission, and the challenge is disentangling the γ-ray lines that belong to each nucleus. Hamilton's group utilized for these measurements Gammasphere, an array of 100 or more Compton-suppressed Ge detectors, normally used for looking at high-multiplicity γ-ray cascades produced in heavy-ion induced reactions, first at the 88-inch cyclotron at Berkeley and then at the Atlas facility at Argonne. Hamilton realized the opportunity to put a ^{252}Cf fission source into Gammasphere for long γ-γ-γ coincidence runs when no accelerator experiments were scheduled. These long runs produced a huge amount of coincidence data that is analyzed by identifying a few known γ-ray transitions in a given pair of fission products, and then using the γ-γ-γ-γ coincidence data to make new assignments to this pair of identified nuclei. This is painstaking analysis that has been ongoing for 30 years since the first publication in 1992, now totaling 80 papers including 39 letter articles. Over 90 collaborators have joined Hamilton in this massive amount of work.

The results of this research have been extensive, including new level schemes for 187 nuclei. The physics learned has been extremely impressive and diverse. The results have included new insights into the fission process itself and new level structures of more than 90 neutron-rich nuclei. Shape evolution from superdeformation in ^{99,101}Y to maximum triaxiality in ^{111,113}Rh has been traced.[11] The first evidence for chiral vibrational doublet bands in ^{106}Mo and ^{110,112}Ru has been published.[12] Perturbed chiral doublet bands in ^{108}Ru and ^{114,116}Pd were proposed.[13] The coexistence of asymmetric and symmetric shapes in ^{144}Ba was demonstrated, which resonates on a longtime Hamilton theme of shape coexistence and represents *Physical Review Letter* #25 in his career, 55 years after the first one.[14]

The sensitivity of this technique was utilized to make new conclusions on the fission process itself, specifically unexpectedly high neutron multiplicities. Hamilton and colleagues identified fission product pairs indicating that sometimes ^{252}Cf fissions with the emission of even 8 to 11 neutrons, instead of the normal two or three. They suggest that such high neutron number evaporation can be an indication of octupole deformation and/or hyperdeformation in the fission products formed.[15]

3.5. *Super-heavy elements*

The fifth phase of Joe Hamilton's career focused on a very different theme but depended on his same talent for building collaborations, making possible a long series of high-profile experiments resulting in the discovery of elements 115, 117, and 118.

This story involves Yuri Oganessian, who in 1989 became the director of the Flerov Laboratory of Nuclear Reactions at the Joint Institute for Nuclear Research in Russia and is now its scientific leader. In 2005 Oganessian visited Vanderbilt, in part to enlist Hamilton's help in obtaining a ^{249}Bk target needed for the next round of heavy element discovery experiments at Dubna. The only economical way to make ^{249}Bk was in a campaign primarily to produce ^{252}Cf at the Oak Ridge reactor (HFIR). For three years, Joe regularly contacted ORNL to learn if a Cf campaign was under way, and finally in August 2008 the answer was yes.

In September 2008, Ed Zganjar (LSU) and I organized a symposium at Vanderbilt University to celebrate Joe's 50 years on the faculty (see Fig. 4). Over lunch, Hamilton, Oganessian, and ORNL's Jim Roberto formed a partnership to obtain ^{249}Bk for a long Dubna experiment. The irradiation and the radiochemical separation of ^{249}Bk was supported by

Fig. 4.   Group of former Hamilton PhD students attending the 2008 conference in honor of Joe Hamilton (shown seventh from the left). Zganjar is fifth from left, Riedinger sixth.

$500,000 from ORNL and $100,000 from Lawrence Livermore National Laboratory, all negotiated and arranged by Joe.

The preparation and the experiment were difficult and painstaking. The High Flux Isotope Reactor at Oak Ridge was used for intense neutron irradiation of Cm and Am targets for approximately 250 days resulting in 22 mg of separated ^{249}Bk. Its halflife is 330 days, long enough to allow the synthesized material to be shipped across the ocean, fabricated into a target, and then used in a 70-day experiment with ^{48}Ca projectiles from the Dubna accelerator. The experiment worked and the group announced the discovery of element 117 based initially on the detection of six events.[16] The compound nucleus formed evaporated three neutrons to form element 117 with atomic mass 294, and four neutrons for 117 mass 293. Each event was identified by the chain of previously known alpha particle emissions among the daughter nuclei, as shown in Fig. 5.

Also identified in this series of Dubna experiments was new element 118, and these two are now the heaviest known elements, with optimism that the discovery of even heavier elements will soon be possible.

The first public announcement of the discovery of element 117 came in a talk that Joe Hamilton delivered at the *International Nuclear Physics* conference in Vancouver in July 2010. In addition, he had a very strong

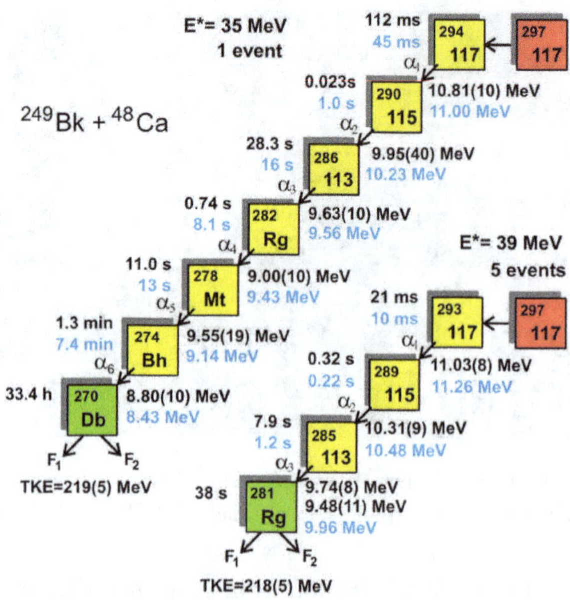

Fig. 5. Observed decay chains originating from isotopes $A = 294$ and $A = 293$ of the new element $Z = 117$. The deduced and predicted lifetimes and alpha-particle energies are shown in black and blue, respectively.

role in naming element 117. In 2016 the International Union of Pure and Applied Chemistry (IUPAC) approved the name Tennessine (Ts) after the state of Tennessee, in view of the critical contributions the state's people and facilities had made.[17] On the periodic chart, element 117 is chemically a halogen, and thus it features the same "ine" suffix of other halogens: fluorine, chlorine, bromine, iodine, and astatine. Element 118 was named Oganesson, in honor of Yuri Oganessian. The decades of former Hamilton students would have been happy with naming element 117 after Joe, perhaps "hamiltine." But, we were not consulted!

4. Final Thoughts

Transitions occur in all fields and must be understood as ideas evolve, priorities change, and key people retire. The retirement of a legend such as Joe Hamilton is certainly a transition to observe. Everyone wonders what will survive and be remembered as each of us retires and eventually expires. In Joe's case, there are four strong answers to the question of what is left behind and what will be remembered.

One legacy comes from the large volume of research in nuclear physics. Joe Hamilton's experiments and ideas on many topics will survive the test of time: shape coexistence, intruder states, reinforcing shell gaps, rotational and vibrational structures, the overall behavior of nuclei far from stability, and the discovery of new elements. Not only is the volume of research output over a career extremely impressive, but also the impact of his results on the broad field of nuclear structure physics is unmatched. His discovery of shape coexistence did not result from a theoretical prediction that a smart experimentalist then found. Rather it resulted from an excellent experimentalist (Joe) measuring something new that led him to forming the idea of shape coexistence.

A second legacy results from his leadership in forming consortia and building instruments that facilitated his research and that of many others. He took the lead on forming and leveraging the first user facility at ORNL (UNISOR), which produced a huge volume of research over four decades, but also set a pattern and established a mode of national lab research that enabled future larger facilities. As ORNL competed for user facilities in the future, the formation and success of UNISOR had a very important impact on winning the competition for a next generation nuclear physics facility (Holifield Heavy Ion Research Facility) and a forefront neutron science facility (Spallation Neutron Source — SNS). The former was built around a 25 MV tandem accelerator (five times the terminal voltage of the EN accelerator used earlier by Hamilton's group) and its coupling to ORIC to make one of the world's best stable beam accelerators for nuclear physics, later upgraded to be our country's first large radioactive-beam accelerator.

The third legacy of Joe's career is his ability to raise funds for his research and the new instruments to be built. He was one of the first science leaders to win *federal* support of research in part by arranging *state* matching dollars. Joe spent his career at a private university, but by its location in Nashville he was able to build personal relationships with important State of Tennessee legislators. His state contacts led to Tennessee support of many projects at ORNL — partial support that was important for leveraging federal support. The first of these was to help purchase in 1971 the mass separator at the heart of UNISOR, then funds to construct a building for the Joint Institute for Heavy Ion Research (JIHIR) at ORNL in 1984, then matching dollars for the Recoil Mass Spectrometer at Holifield in 1990, in addition to state help on other nuclear physics projects in Oak Ridge. Since Vanderbilt is a private university, these state funds were usually appropriated to the UT, and I became Joe's partner in obtaining and then administering these funds for Oak Ridge based initiatives.

Long-term benefit to science in Tennessee resulted from the activism that Joe showed in working with state government. Though Tennessee is not a rich state (it has no state income tax), the state has invested regularly in Oak Ridge based science projects. The UT teamed with Battelle Memorial Institute to form UT-Battelle LLC that won the contract to manage ORNL in 2000 and has maintained that role ever since. Part of that winning proposal was the State of Tennessee's commitment to provide $12 million to build three new UT-ORNL joint institutes based on our successful 1984 model — Hamilton inspired — of the first joint institute (JIHIR), funded and operated by UT, Vanderbilt, and ORNL.

The fourth surviving legacy of Joe Hamilton's career is his impact on people. A parent works to raise children in the right way, instill correct values, and follow and support them as they progress in their lives and careers. Joe Hamilton has done all of these things for his students. Around 90 of us received our Vanderbilt graduate degrees by working on research in nuclear physics in Hamilton's group. These graduates have proceeded to successful careers in various fields, some trying to closely follow his footsteps in nuclear physics (me included). We all benefited from his example — hard work, long hours, care for detail, optimism for the challenge at hand, and the inability to accept no for an answer. None of us can match him in any of these ways, but we each inherit some aspects of each. Like a parent, Joe has stayed involved in our progress, our challenges, and our successes. He has always shown interest in each of our careers, helped us succeed, and proposed us for awards when we do. Each of us values the continuing contact and support as much as the training he gave us in graduate school. I feel this perhaps more strongly than some others, as I continued to work closely with Joe throughout my long career and benefited every step of the way. He has been my mentor, my role model, my colleague. As I wind down my career, I feel very fortunate to have worked so closely and for so long with a legend. Few people have such a sterling opportunity and benefit in such a profound way. I join many others in thanking Joe for all he has done.

References

1. W. G. Smith, J. H. Hamilton, R. L. Robinson, and L. M. Langer, Beta Decay of ^{161}Tb, *Phys. Rev.* **104**, 1020–1025 (1956).
2. L. L. Riedinger, N. R. Johnson, and J. H. Hamilton, Band Mixing in ^{152}Sm and ^{154}Gd, *Phys. Rev. Lett.* **19**, 1243–1247 (1967).
3. L. L. Riedinger, N. R. Johnson, and J. H. Hamilton, β- and γ-Vibrational Bands of ^{152}Sm and ^{154}Gd, *Phys. Rev.* **179**, 1214–1229 (1969).
4. J. H. Hamilton, A. V. Ramayya, and L. C. Whitlock, Further Test of M1 Admixtures in the Decay of 2^+ Beta Vibrational States, *Phys. Rev. Lett.* **22**, 65–67 (1969).
5. L. Varnell, J. H. Hamilton, and R. L. Robinson, Collective States in ^{178}Hf via Coulomb Excitation, *Phys. Rev.* **C3**, 1265–1268 (1971).
6. R. M. Ronningen, F. Todd Baker, Alan Scott, T. H. Kruse, R. Suchannek, W. Savin, and J. H. Hamilton, Evidence for the Nonrotational Interpretation of <0+‖M(E4)‖4+> in ^{180}Hf, *Phys. Rev. Lett.* **40**, 364–367 (1978).
7. J. H. Hamilton, *et al.*, Crossing of Near-Spherical and Deformed Bands in ^{186,188}Hg and New Isotopes ^{186,188}Tl, *Phys. Rev. Lett.* **35**, 562–565 (1975).
8. J. H. Hamilton, *et al.*, Evidence for Coexistence of Spherical and Deformed Shapes in ^{72}Se, *Phys. Rev. Lett.* **32**, 239–242 (1974).
9. J. H. Hamilton, *et al.*, Lifetime Measurements to Test the Coexistence of Spherical and Deformed Shapes in ^{72}Se, *Phys. Rev. Lett.* **36**, 340–342 (1976).
10. R. B. Piercey, *et al.*, Evidence for Deformed Ground States in Light Kr Isotopes, *Phys. Rev. Lett.* **47**, 1514–1517 (1981).
11. J. H. Hamilton, *et al.*, Super deformation to maximum triaxiality in A = 100 – 112; superdeformation, chiral bands and wobbling motion, *Nucl. Phys. A*, **834**, 28c–31c (2010).
12. S. J. Zhu, *et al.*, Triaxiality, chiral bands and gamma vibrations in A = 99–114 nuclei, *Progress in Particle and Nuclear Physics*, **59**, 329–336 (2007).
13. K. Butler-Moore, *et al.*, High-spin States in Neutron-Rich Even-Even Pd Isotopes, *J. Phy. G* **25**, 2253–2269 (1999).

14. S. J. Zhu, *et al.*, Coexistence of Reflection Asymmetric and Symmetric Shapes in ^{144}Ba, *Phys. Rev. Lett.* **124**, 032501 (2020).

15. B. M. Musangu *et al.*, Anomalous neutron yields confirmed for Ba-Mo and newly observed for Ce-Zr from spontaneous fission of ^{252}Cf, *Phys. Rev. C* **101**, 034610 (2020).

16. Yu. Ts. Oganessian, *et al.*, Synthesis of a New Element with Atomic Number Z = 117, *Phys. Rev. Lett.* **104**, 142502 (2010).

17. IUPAC Is Naming The Four New Elements Nihonium, Moscovium, Tennessine, and Oganesson, *IUPAC*, August 6 (2016).

Joe Hamilton-led Innovations: The University Isotope Separator — Oak Ridge and the Joint Institute for Heavy Ion Research

Ed Zganjar

*Alumni Professor Emeritus,
Louisiana State University, USA*

Joe Hamilton's conception and bringing into being the University Isotope Separator in Oak Ridge (UNISOR) and the Joint Institute for Heavy Ion Research (JIHIR) in Oak Ridge and their impacts on new facilities and science are described.

1. Introduction

The first university collaboration for low-energy nuclear research at a Department of Energy National Laboratory was initiated by Joe Hamilton in the late 1960s called UNISOR for University Isotope Separator at Oak Ridge. The timeline for UNISOR is given in Table 1.

In 1968, ORNL developed a proposal for a new heavy ion accelerator with a university user's group in the proposal. Joe was asked to form a university user's group. They met at Vanderbilt to develop a section of the proposal. With funds for new projects very tight in Washington, the project did not get funded.

Table 1.   Timeline of UNISOR.

- **1968:** Oak Ridge National Laboratory developed a proposal for a new heavy ion research facility to include participation of university scientists. A university group was formed with Joe Hamilton as chair.
- **1969:** When this was not funded, the idea of UNISOR was conceived by Joe. At a meeting of University scientists at Vanderbilt, chaired by Joe, an organization was formed to purchase and place an isotope separator on line to the ORNL cyclotron with an on-site university staff to maintain the facility. Joe immediately set out to raise major funds for the separator and its operation.
- **1971:** The formal organization of UNISOR was made with Oak Ridge Associated Universities serving as the physical agent.
- **1972:** The Isotope separator, built in Denmark, was delivered to ORNL.
- **1972:** On June 19, the formal dedication of UNISOR was held.
- **1972:** On September 2, the first on line experiment was made.

With the drying up of funds for small on-campus nuclear research, Joe then conceived the idea of a university user's group to use the existing ORNL cyclotron. They would purchase a magnetic isotope separator, put it on line with the cyclotron, hire a staff to maintain the facility, and work with the university scientists in research on new nuclei well removed from the stable ones in nature. This would give the university scientists forefront research opportunities. Since no new projects were being funded, they would have to raise the funds to buy the separator and provide the order of 50% of the operating budget including for the on-site staff. Joe first went to his administration to ask for $15,000 for the separator and $10,000 per year for five years for operations. They agreed if he could raise the other needed funds. Next, he approached Bill Bugg, chair of physics at UT, and he obtained an agreement to match VU. It was harder to convince other Deans in out-of-state universities to put money into ORNL. Joe's convincing argument was if you want your university nuclear scientists to be able to do forefront research, this would be their only option. He raised $90,000 for the separator and over $50,000 a year for five years for operating expenses. Then came a windfall for equipment. Joe was a member of a group at his church who met for dinner and to discuss issues of concern to them once a month. At one meeting, Joe discussed his efforts to develop the UNISOR project but he needed more funds. In the group was Tennessee Governor Ellington's Chief Executive Officer Bo Roberts who said to Joe this sounds like something the State

Fig. 1.   The founding members of UNISOR meet in ORNL. These were U. Alabama Birmingham, Emory U., Furman U., Georgia Institute of Tech., U. Kentucky, Louisiana State U., U. Massachusetts, U. South Carolina, U. Tennessee Knoxville, Tennessee Technological Inst., Vanderbilt U., Virginia Polytechnic Inst., State U., and ORAU.

can put some money into. Send me a proposal to match the $90,000 you have for equipment. Joe presented a request to the State committee chaired by Mr. Roberts and they awarded another $90,000 for the project. Mr. Roberts said to Joe that governments at every level, city, state, and federal level receive more requests for funds than they can award. In general, they fund those where they know the person and trust what he proposes is important and that he can do what he says. This insight was important in developing the Joint Institute described later.

UNISOR was the first nuclear collaboration of scientists with funding from universities, a State government, a national laboratory, and a federal energy agency.

In 1972, the magnetic isotope separator, built in Denmark, was delivered to ORNL. It was the first such separator put directly on line to a heavy ion cyclotron to study the structure of nuclei far from the stable ones in nature in the world. The formal dedication of the facility was held in September 1972 with a ribbon-cutting ceremony attended by all of the scientists involved as seen in Fig. 2.

At the dedication, Alvin Weinberg, Director of ORNL said "In the days of austerity in the physical sciences, it's a particular pleasure to be able to point to a project that goes counter to that trend. UNISOR is remarkable not only for its sheer existencebut also for the extraordinary

Fig. 2. The formal dedication of the UNISOR was held in September 1972.

Fig. 3. The UNISOR separator and collection boxes being installed by the on-site staff.

measure of cooperation between so many institutions that it represents". The unique UNISOR consortium was described by Joe in Science 185, 4154. September 1974.

Figure 3 shows the separator and collection boxes being installed by the on-site UNISOR staff. Over the years, as faculty retired, some universities dropped out and nine others joined including U. Maryland, Michigan State U., Mississippi State U., Mississippi College, U. Notre Dame, Oregon State U., Oxford U., UK, Rutgers U., and Texas A & M U.

As time went on, new facilities were added to UNISOR. In 1981, a co-linear laser facility was put on-line, as shown in Fig. 4, funded by the universities. By laser excitation of an atomic transition in the isotope separator beam line, hyperfine structures are measured as well as the isotope shifts of the transition. From these data, the magnetic dipole moments, spectroscopic quadrupole moments, and isotopic changes in mean-square charge radii can be deduced. Also, precise measurements of atomic lifetimes can be measured.

In 1988, UNISOR put on line a ^{3}He-^{4}He dilution refrigerator to measure angular correlations of nuclear radiations to determine M1/E2 mixing ratios and nuclear spins. When combined with electron conversion measurements, one can obtain the exact fraction of the electric monopole component in transitions between levels of the same spin and parity. The UNISOR universities provided more than half of the $250,000 for the refrigerator. Figure 5 shows the celebration of the new facility, the first

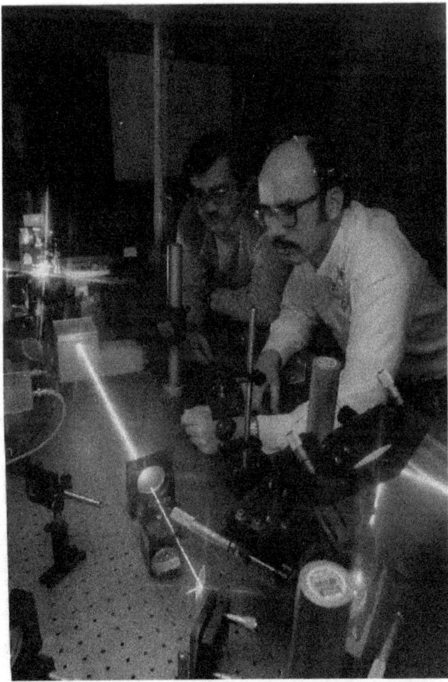

Fig. 4.   In 1981, a co-linear laser facility, funded by the universities, was put on line.

Fig. 5.   Celebration of the new UNISOR facility.

such in the US, by all the students, post doctoral fellows, staff, faculty, and ORNL scientists involved in the project.

In the 1980s, Professor Ramayya and Joe joined in research with Prof. Cormier, U. Rochester, who had built the first on line isotope separator in the US to study the structures of isolated nuclei produced in a nuclear reaction. They became excited about this new way of studying, particularly nuclei. Joe convinced Cormier and Jerry Cole, a former PhD student at Idaho National Engineering Lab, to design a new generation recoil mass spectrometer with Joe's specifications to do inverse reactions of a light ion on a heavy target as well as heavy ions. It would be a cooperation of UNISOR and ORNL. The project would cost 4–6 million dollars including a new building and beam lines. So, Joe raised first $150,000 from Vanderbilt, then similar amounts from U. Tennessee, UNISOR, ORAU, and INEL. Nevertheless, DOE turned down this proposal choosing to fund a million dollar much smaller old design RMS at another laboratory. He then convinced the Deputy Director of ORNL to provide $200,000 of non-DOE funds for the project (one ORNL scientist said they never knew such funds existed and asked what motel he photographed him coming out of jokingly). Then having cultivated Senator Henry, Chair of the Tennessee

Senate finance committee, following Mr Robert's insight, he went to Senator Henry to request $500,000 from the State for the RMS. Senator Henry arranged a meeting with him, Joe, a UT representative, Representative Bragg, Governor McWerter, and his Finance Director for Joe to present his proposal. At his conclusion, the Governor asked how much is Vanderbilt putting into this project, and Joe said $150,000. Then he asked how much is UT putting in, and Riedinger said $150,000. To which the governor replied, "where are they getting the money, to hear them talk they do not have two dimes to rub together?" After this, the Governor looked at his Finance Director and asked "If I sign off on this, are you going to give me trouble?", to which the Director said, "The last time I looked you were the Governor". He responded, "Let's do it, they need jobs over there!". Now, Joe had raised over $1,500,000 for the project. DOE then called Joe and said cease and desist you can have the balance of funds you need. To which Joe replied, "I just saved you because our RMS will do important experiments the other lab cannot do". As examples, experiments involving proton and alpha decays out to the proton unbound drip line were studied.

Figure 6 shows the world's finest RMS nearing completion with Joe, his post doc JJ Das, his PhD graduate student Tom Gintner (now a staff scientist at FRIB at Michigan State), and UNISOR staff Paul Mantica.

Fig. 6. The world's finest Recoil Mass Separator nearing completion at Oak Ridge.

2. The Holifield Heavy Ion Research Facility HHIRF, later Holifield Radioactive Ion-Beam Facility HRIBF (UNISOR > UNIRIB)

When DOE decided to build a new more versatile heavy ion accelerator research, the most economical and best facility was to construct the world's largest tandem accelerator and couple it to the ORNL cyclotron as an injector. DOE wanted this facility to be different from the previous accelerator facilities at their National Labs which were primarily used only by in-house staff scientists. It was to be a nationally and internationally used facility. A National Users Policy Committee was formed with scientists from two national labs and three universities, including Joe (who served two terms as chair), to oversee that the facility would be nationally used.

At one of their early meetings, Allan Bromley, first Chair of the committee, said to Joe in private that his UNISOR project was essential in getting ORNL selected as the site of the new facility. He added in 1971 that Joe submitted to DOE a proposal for UNISOR to use the ORNL cyclotron for five years using a new isotope separator purchased by them and pledging half of the operating funds for five years. What Joe did not know, Bromley had just chaired a committee that had given at the same time to Glenn Seaborg, Chair of the Atomic Energy Commission, their report recommending that the old ORNL cyclotron be closed. Seaborg decided to fund the UNISOR project and give the ORNL cyclotron five more years. If he had accepted our report to close the cyclotron, then it would not have been available in 1974 to reduce significantly the cost of a new facility at ORNL, so Holifield would not be built. Figure 7 shows the outside of the Holifield facility.

Around 1992, DOE considered closing the ORNL accelerator facility. Joe and Lee Riedinger, UT, made a visit to House of Representatives Marilyn Lloyd to ask her help to keep Holifield open and convert it into the first US radioactive ion beam facility. That same afternoon, she went to visit the nuclear division of DOE and asked them to convert Holifield to an RIB facility. The next morning, ORNL got a request from Washington for a proposal to convert their accelerator to an RIB facility which they then funded (described by Bertrand Physics Division Director at the 2008

Fig. 7.   The Holifield facility.

conference at Vanderbilt for Joe). It turns out this was an important step because the UNISOR separator was used as a test facility to develop radioactive ions for accelerator beams. UNISOR changed its name to UNIRIB. HRIBF was very successful in showing the power of using radioactive ion beams to study nuclear matter.

3. The Joint Institute for Heavy Ion Research

As the new committee met to help develop Holifield into a nationally used facility, Joe saw the need to have a user's building on site with bedrooms, a lounge, and a kitchen where outside users could stay and hold meetings to discuss their progress. The city of Oak Ridge was ten miles away. If something happened and they needed the expertise of a group member who was sleeping 10 miles away, this could be a serious delay problem.

First, Joe had to raise the money to build such an institute. He asked his Dean for $100,000 to initiate the project and would ask U. Tennessee and ORNL to join the formation of an institute which would be managed by them. Going up the chain command at VU, the Provost said they do not have such funds to put in a building in Oak Ridge. They can rent Joe a hotel when he needs a room. He suggested meeting with Chancellor Heard to consider Joe's request. The Chancellor said it is true they do not normally have such funds, but "the opportunity for national leadership is so great that if you can raise the other needed funds, I will personally find the money". Joe convinced UT to also provide $100,000 and ORNL to provide $312,000 to complete building one of the institutes.

However, there remained a serious roadblock. DOE did not allow universities to have buildings on DOE land. A meeting was held with the local DOE managers in Oak Ridge. As each objection to a building on

Table 2.   A timeline of the construction of the three Joint Institute buildings.

- **1980:** Construction of the first building with sleeping rooms, a lounge, a dining, and a kitchen at a cost of $512,000 provided by Vanderbilt, U. Tennessee, and ORNL.
- **1982:** A land lease agreement was signed to allow the universities, the State of Tennessee, and ORNL to construct and manage a building on DOE land.
- **1982:** A Memorandum of understanding was signed by ORNL, UT, and Vanderbilt to define the operation of the institute by a Policy Council with each having a representative. UT would manage the operating funds. At the second meeting of the Council, Joe was elected Chair.
- **1983:** Joe wanted the Institute to be a conference center as well. He got Chancellor Heard to invite Governor Alexander, a VU graduate, to a breakfast where he asked the State to provide $350.000 to go with $150.000 from the universities to fund Joe's idea of making the Institute into a conference center with a lecture hall, secretarial space, offices for VU and UT faculty, a home for UNISOR staff (which they would pay for) and visitors, and some lab space. The Governor agreed and construction soon began on the second building.
- **1984:** A dedication ceremony and conference was held in October. With the concurrence of the UT and VU Chancellors, Joe requested funds from the State of Tennessee "to create the Institute as a Center of Excellence in Higher education, the likes of which the State had never seen". The Governor did not reply. In 1985 January, the Governor in his State of the State address proposed putting twenty million dollars into Centers of Excellence in Higher Education across the state. Chancellor Reese of UT said to Joe "it is remarkable how closely the Governor's Centers of Excellence in Higher Education parallels your letter to him for funds to make our institute a Center of Excellence".
- **2002:** Joe requested, through Senator Henry, that the State provide $300,000 with additional university funds for a third wing of the Institute with offices for the rapidly expanding theoretical work at the institute. This was done.

Fig. 8. JIHIR Dedication in 1984 with Nobel Laureate Glenn Seaborg the speaker and with Representative Lloyd, the Chancellors of VU and UT, the President of the U. Tennessee system, Director of ORNL, Joe, and local and Washington representatives of DOE.

Fig. 9. The Policy Council members for the Joint Institute.

DOE land was put forward, it was shot down, until after all objections were shot down they said "Oh build that Hamilton Hilton".

When the Centers of Excellence were funded, the Science Alliance at UT received the largest funding and the Joint Institute received the

Fig. 10.   The three buildings of the Joint Institute for Heavy Ion Research.

Table 3.   The major accomplishments of the institute.

- Hosted more than 1,300 research visitors.
- Supported hundreds of students and young faculty.
- Contributed to significant advances in nuclear physics both experiment and theory.
- Inspired and contributed to major instrument development.
- Led the development of the theory of nuclear structure far from stability.
- Sponsored more than 100 international conferences and workshops attended by over 4,500 scientists.
- Leveraged talent and resources from multiple institutions to sustain a vibrant scientific community.
- Support for more than 300 long-term visitors in experiment and theory.
- Short-term visits by many internationally distinguished nuclear theorists.

funding it had requested to bring visiting scientists from all over the country and the world to work for periods of a few weeks to a year or more with UT, VU, and ORNL scientists and to have a secretary and funds to host international conferences. Indeed, it did become a major Center of Excellence in Tennessee.

For his many contributions to facilities and research at ORNL, Joe was appointed the first Visiting Distinguished Laboratory Fellow

JIHIR has Pioneered the joint institute model

- Welcomed more than 1300 research visitors
- Supported hundreds of students and young faculty, many now leaders in nuclear science
- Contributed to significant advances in nuclear physics experiment and theory
- Inspired and contributed to major instrument development
- Led the development of the theory of nuclear structure far from stability
- Sponsored more than 100 conferences and workshops attended by more than 4500 people
- Leveraged talent and resources from multiple institutions to sustain a vibrant scientific community
- Support of more than 300 long-term visitors in experiment and theory
- Short-term visits by many international distinguished nuclear theorists

Fig. 11.

JIHIR has also played a pivotal role in the transformation of ORNL

- **ORNL's first major user facility**
  - We now have SNS, Leadership Computing, the Center for Nanophase Materials, and many others serving thousands of university users annually

- **The first Joint Institute**
  - Paving the way for joint institutes in computational sciences (JICS), biological sciences (JIBS), neutrons sciences (JINS), and advanced materials (JIAM)
  - Creating a vehicle for $200M in new shared programs in high-performance computing (NSF petascale grant) and bioenergy (DOE Bioenergy Science Center)

- **A compelling demonstration of the power of Laboratory-University collaboration**

Fig. 12.

Fig. 13. Thanks to Joe Hamilton for his vision and leadership in making UNISOR and JIHIR so successful.

at ORNL. Laboratory Fellow is the highest recognition a scientist can receive at ORNL.

In 2008, the Physics Department at Vanderbilt organized a small Symposium in honor of Joe's 50 years of teaching and research at Vanderbilt. Jim Roberto, then Deputy Director of ORNL, and Fred Bertrand Chair of the Physics Division both spoke about all of Joe's contributions and the impact they have had on ORNL.

Figures 11 and 12 illustrate the impact of the JIHIR at ORNL. Published with the kind permission of Jim Roberto.

Summarizing Joe's institute contributions: first, UNISOR which had a critical contribution to the Holifield Heavy Ion Lab being placed at ORNL, which led Joe to develop the Joint Institute for Heavy Ion Research and which paved the way for the State of Tennessee to give $16,000,000 for building three new institutes at ORNL that have literally transformed ORNL.

Thank you Joe for being my major professor, helping my career along, as you have done. Thank you, Joe.

Joe Hamilton's Support of Science at the ORNL Physics Division

Fred Bertrand

Physics Division, Oak Ridge National Laboratory, USA

This chapter is taken from the 2008 contribution to the symposium celebrating J. H. Hamilton's 50 years of teaching and research at Vanderbilt University.

1. Some Contributions of Joe Hamilton to the ORNL Physics Division

1972: "In the days of austerity in the physical sciences, it's a particular pleasure to be able to point to a project that goes counter to that trend. UNISOR is remarkable, not only for its sheer existence, but also for the extraordinary measure of cooperation between so many institutions that it represents." Alvin M. Weinberg, Director, Oak Ridge National Laboratory, at the UNISOR dedications, June 19, 1972.

1973: One of the three speakers (P. Stelson and C. Moak) at the meeting with Glenn Seaborg, chairman AEC (DOE), to present the ORNL proposal for a new heavy ion accelerator combining ORIC and a proposed new tandem. Spoke on behalf of university users of the facility.

1974–1982: Served on the National Policy Board for the new Holifield Heavy Ion Research Facility. The committee was charged to ensure this facility would be the first nationally used facility in nuclear physics at a national laboratory. He chaired the committee in 1975 and 1979.

1980–1981: Convinced Vanderbilt to contribute first $100,000 to initiate the Joint Institute for Heavy Ion Research to help make the new Holifield Heavy Ion Research Facility a major world center in research. University of Tennessee then contributed $100,000 and ORNL $312,000 for the Joint Institute's first building. Hamilton has been on the three-member policy council that oversees the Joint Institute since its inception and has served as Director of the Institute through the present except for three years in the 1980s.

At the dedication of the Holifield Heavy Ion Research Facility, Edward A. Friedman, Director, Office of Energy Research, U.S. Department of Energy, said, "In basic nuclear research, we have two outstanding success stories — the University Isotope Separator at Oak Ridge (UNISOR) and the Joint Institute for Heavy Ion Research."

1984: J. H. Hamilton proposed in a letter to Governor Alexander that the State of Tennessee fund the Joint Institute, including a visitor's program to make it a "Center of Excellence in Higher Education in the State of Tennessee." The next year Governor Alexander proposed and the legislature funded the new $20,000,000 Centers of Excellence Program for the State of Tennessee. The University of Tennessee was the largest benefactor of such funds for the Science Alliance with ORNL. In addition to funding and supporting new Distinguished Scientist positions, the Alliance included $175,000 annually for the JIHIR, with Vanderbilt University contributing $12,500 annually to the Institute and DOE

$60,000 to the Institute. After the Governor initiated the Centers of Excellence Program, the Chancellor of the University of Tennessee remarked how strikingly similar Governor Alexander's title Centers of Excellence and justification were to those in the letter J. Hamilton had written the Governor the year before.

1986: Based on his experience with the University of Rochester recoil mass spectrometer, the first such device built for in-beam gamma-ray spectroscopy, he established a group to design a new-generation RMS that could do inverse reactions. He then raised $150,000 each from Vanderbilt, University of Tennessee, UNISOR, ORAU, and INEEL. When DOE turned down the Vanderbilt-ORNL proposal in 1989 because of its cost, including building and beam lines, of about $4,000,000 in favor of a smaller, and much less costly, old-design RMS in existing space at Argonne, he then raised an additional $500,000 from the State of Tennessee and $300,000 of non-DOE funds from the ORNL's Director's office. He raised in total about $1,560,000 for the RMS from non-DOE sources. Then, DOE approved the project. The procurement and construction at Danfysik of the RMS were overseen by R. L. Robinson (ORNL), J. D. Cole (Idaho), and J. H. Hamilton. When the RMS components arrived, a Vanderbilt/Joint Institute postdoctoral fellow with RMS assembly experience was hired by him, and a Vanderbilt graduate student assigned to help in the assembly and testing of the facility (T. Ginter). This RMS is the centerpiece of current nuclear structure work at Holifield and is arguably the world's finest RMS.

~1991: J. H. Hamilton made major lobbying efforts in Washington on behalf of the proposal to make Holifield a Radioactive Ion Beam facility and was helpful in getting UNISOR universities to lobby on ORNL's behalf. Of particular importance were his and L. L. Riedinger's contacts with Representative Marilyn Lloyd, who personally went to see the DOE officials to make Holifield an RIB facility. They immediately acted following her visit.

~1996: When the University of Tennessee sought State support for $7,000,000 for an Institute of Neutron Science to help DOE decide to move ahead on the $1,300,000,000 Spallation Neutron Source, the sale was easy according to UT officials because State officials were well aware of the success story of the Joint Institute of Heavy Ion Research.

1997: Vanderbilt and the University of Tennessee provided on the order of $150,000 each that helped convince DOE to provide the remaining funds for the new $2,000,000 Clover Ge detector array for the RMS.

JHH established "User" Science as a required mode of operation at DOE NP Accelerator facilities. **The original outside user!**

Deeply committed to excellence in science.

Continually supported ORNL Physics Division programs.

Without his efforts, many of our most successful projects would not have gotten off the ground.

He has brought in millions of dollars of equipment to the Holifield facility and has provided many excellent students to work at the Facility.

Joe is a friend and colleague to all of us.

Joe does not know the meaning of "can't do". "No" is not an answer he is willing to accept.

2. Visiting Distinguished Laboratory Fellow Program

Guidelines: The Visiting Distinguished Laboratory Fellow Program recognizes distinguished visitors from universities, industry, and other institutions for extraordinary contributions to ORNL through scientific collaboration, public outreach, and leadership in interactions with the scientific community and other stakeholders at the federal, state, and local levels. This position complements the Visiting Distinguished Scientist position, which brings in outstanding individuals on site for the purpose of scientific collaboration. The Visiting Distinguished Laboratory Fellow Program celebrates broader contributions to the Laboratory across a broad spectrum of scientific and programmatic activities.

Qualifications: Visiting Distinguished Laboratory Fellows should be outstanding scientists with a record of leadership in the scientific community. They should have a demonstrated long-term commitment to ORNL through prolonged scientific collaboration with Laboratory staff and a sustained record of mentoring ORNL programs with significant impact at the federal, state, and local levels. Typically, these candidates may have developed a field of science or promoted a new program within the Laboratory. Additionally, they may have broadened the societal impact of Laboratory programs as conveyors of science in public outreach programs.

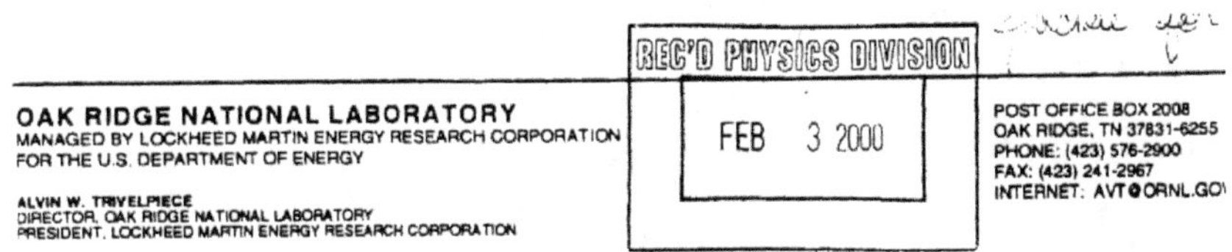

OAK RIDGE NATIONAL LABORATORY
MANAGED BY LOCKHEED MARTIN ENERGY RESEARCH CORPORATION
FOR THE U.S. DEPARTMENT OF ENERGY

ALVIN W. TRIVELPIECE
DIRECTOR, OAK RIDGE NATIONAL LABORATORY
PRESIDENT, LOCKHEED MARTIN ENERGY RESEARCH CORPORATION

REC'D PHYSICS DIVISION

FEB   3 2000

POST OFFICE BOX 2008
OAK RIDGE, TN 37831-6255
PHONE: (423) 576-2900
FAX: (423) 241-2967
INTERNET: AVT@ORNL.GOV

February 1, 2000

Professor Joseph H. Hamilton
Landon C. Garland Distinguished Professor of Physics
Vanderbilt University
Box 1807, Station B
Nashville, Tennessee 37235

Dear Professor Hamilton:

It is with pleasure that I announce your selection as Oak Ridge National Laboratory's first Visiting Distinguished Laboratory Fellow. The Visiting Distinguished Laboratory Fellow program was established to recognize scientists from universities, industry, and other institutions for extraordinary contributions to Oak Ridge National Laboratory (ORNL) through sustained leadership in scientific and programmatic activities.

For more than 30 years, you have been a powerful force for nuclear physics at ORNL. Your contributions have included leadership in the creation of the University Isotope Separator at Oak Ridge (UNISOR) consortium in 1971 and the Joint Institute for Heavy Ion Research (JIHIR) in 1981. Your dedication to laboratory-university cooperation and your effectiveness in advancing collaborative scientific projects at ORNL have been essential to the success of our nuclear physics effort. These contributions have been complemented by a long history of highly productive interactions with you and your students.

As a Visiting Distinguished Laboratory Fellow, we hope that you will continue your close involvement with ORNL. With your concurrence, Dr. Fred Bertrand, director of our Physics Division, will be contacting you to extend an invitation for you to present a lecture of your choosing at ORNL. At this event, you will receive a plaque and small honorarium.

Your appointment sets the standard for future Visiting Distinguished Laboratory Fellows at ORNL. We thank you for your commitment and energy in support of the Laboratory, and we look forward to your continued friendship and collaboration.

Sincerely,

Alvin W. Trivelpiece

Joseph H. Hamilton Oak Ridge National Laboratory First Ever Visiting Distinguished Laboratory Fellow!

Thank you Joe for all you have done for ORNL and especially for your unfailing support of the Physics Division.

Let there be many more years!!

Prof. J. H. Hamilton and My Research

S. Frauendorf

*Prof. Emeritus, Department of Physics and Astronomy,
University Notre Dame, Notre Dame, IN 46556, USA
sfrauend@nd.edu*

1. Shape Coexistence: Two Sides — One Topic

This special volume is devoted to Prof. Joseph Hamilton's retirement from his long impressive research and teaching carrier. Here, I present some words of appreciation from my personal perspective. I retired two years before Joe. During a large part of my scientific carrier, I have been linked with Joe and his group in one way or the other. It all started early on in 1971. Right after defending my PhD at the Technical University Dresden, I left for a post doc appointment at the Joint Institute for Nuclear Research (JINR) Dubna in the Soviet Union. Doing research at the hub of the nuclear science in the East Block was a great opportunity for a young East German. My thesis work had been on nuclear rotation, so it would have been natural to join Dr. Igor Michailov's group working on this subject. However, Prof. Vadim Soloviev, the head of the theory department, suggested to change subject and collaborate with Dr. Vitali Pashkevich on shape and fission isomerism. It was a hot topic at the time. Vilen Strutinsky had just published his famous shell correction method,[1,2] which quantitatively described the fission isomers for first time and predicted the existence of superheavy elements. So, I started working with Vitali, who had good connections with Strutinsky. We became good friends and kept collaborating till the end of his life. We set out studying the proton-rich Hg isotopes because the then recent measurements of hyperfine splitting at ISOLDe in CERN showed a large difference in the charge radius between the even-A and odd-A isotopes, which could be attributed to the coexistence of a weakly and a strongly deformed shape.[3,4] The calculations got somewhat slowly going because I wanted to finish and publish the left — overs from my thesis.

With his pioneering work on the coexistence of spherical and deformed shape in ^{72}Se, Joe and his group opened up the field of shape coexistence[5] together with the Chalk River[10] and Berkeley[11] groups. Figure 1 illustrates the seminal study, which is cited 224 times. The spectrum in the upper panel shows the spherical ground state 0_1^+ and the spherical vibration states 2_1^+ and 4_1^+. In addition, there are the deformed state 0_2^+ and the deformed rotational state 2_2^+. The lower panel displays the square of the rotational frequency $\omega(I)$ and the moment of inertia $\mathcal{J}$ for the γ transitions $I \rightarrow I - 2$ between the lowest states of given I. The frequency is constant for the $2_1^+ \rightarrow 0_1^+$ and $4_1^+ \rightarrow 2_1^+$ as expected for vibrational excitations and then changes into the sloped line when the rotational states based on the deformed 0_2^+ become the lowest. In a subsequent lifetime measurement, Joe's group confirmed the small deformation of the vibrational states and the large deformation of the rotational ones.[6]

In 1972, I attended the International School on Nuclear Structure organized by the JINR in Alushta, a vacation resort on Crimea peninsula. During his long carrier, Joe put much effort in establishing and maintaining contacts with the scientists from the other side of the iron curtain and with Russia when the curtain fell and the USSR was gone. This outreach to the East culminated in the Vanderbilt–Oak Ridge–Dubna joint search for super heavy elements resulting the discovery of Tennessine. The Alushta school, where Joe presented two lectures, marks the beginning his relation building. The school offered ample opportunities to speak with the lecturers beyond the official schedule, which was my first personal encounter with Joe. Vitaly and I to discuss with him our calculations of shape coexistence in the Hg isotopes. It turned out that he was investigating the phenomenon in the rotational bands of ^{186,188}Hg[6] and that similar work on ^{184,186}Hg was going on in the Berkeley and Chalk River Laboratories.[10,11]

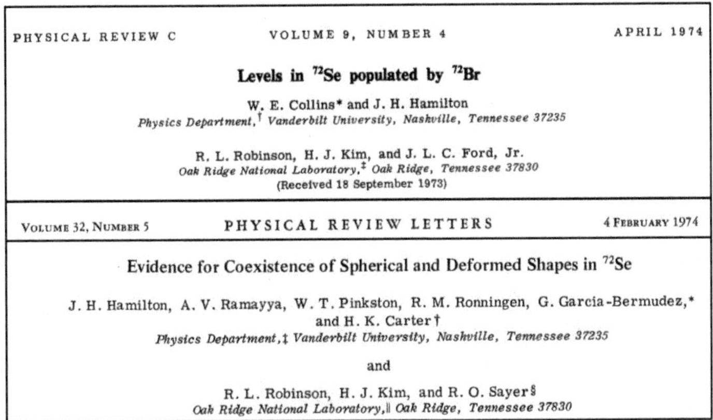

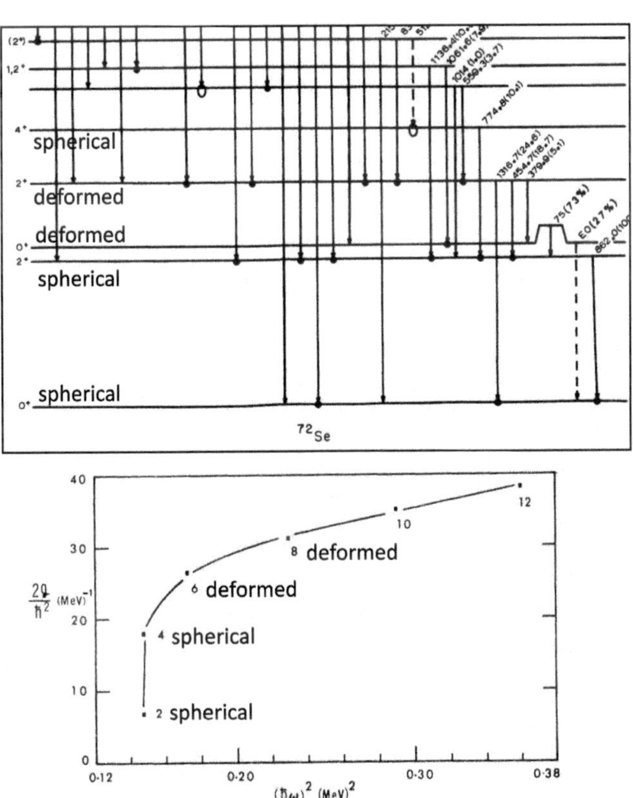

Fig. 1. Upper panel: Energies of ^{72}Se. The various states of the nucleus are labeled by I^π, where the spin I indicates how rapidly the nucleus is rotating and the parity π the symmetry of the states. In the text, they are labeled by I_n^π, where n count the states with the same I and π. Lowest panel: Moment of inertia $\mathcal{J} = \mathcal{I}/\omega(\mathcal{I})$ as a function of the square of the rotational frequency $\omega(I)$, which is equal to one-half of the energy difference between the lowest states with $I - 2$ and I. From Ref. 5.

It was not the time to start an East-West collaboration. However, inspired by the discussion, I pushed the project when we were back to Dubna. The results were published in Ref.[8] and used by Joe and his collaborators to interpret their experiment[9] on ^{186,188}Hg. Figure 2 shows the calculated energies of the Hg isotopes as functions of the deformation parameter ε, which measures how strongly the nuclear shape deviates from a sphere. In all cases, there is a minimum for negative ε, which corresponds to a weakly deformed oblate (pancake-like) shape, and a minimum for positive ε, which corresponds to a strongly deformed prolate (cigar-like) shape. The nucleus settles at the minima, that is, the two shapes coexsist in one of the same nucleus as states with somewhat different energy. The upper left panel illustrates how relative energy of the two shapes changes when the nucleus rotates, where the spin I indicates how rapidly. For ^{184}Hg, the oblate minimum is lower than the prolate one for the spins $I = 0, 2$. The order changes at $I = 4$ because the moment of inertia is larger for the prolate shape. The calculated deformation energies for ^{186,188}Hg look similar. The lower panel demonstrates that the resulting crossing of the two sequences of states with increasing I (rotational bands) is seen in the experimental energies measured by Joe and his collaborators.[9] The experiments on ^{184,186}Hg by the authors of Refs.[10,11] show the crossing bands as well. In addition, they confirmed their respective weak and large deformations by lifetime measurements. As seen in the upper right panel, the ground states of ^{181,183,185}Hg are found to be strongly prolate in contrast to the weakly oblate ground states of their neighbors ^{182,184,186}Hg, while the ground states of ^{187,189}Hg are found to be weakly oblate like their neighbors ^{186,188,190}Hg. This explains the differences in the charge radii of the Hg isotopes measured by the authors of Refs. 3 and 4.

After my return to the Central Institute for Nuclear Research in Dresden in 1974, Vitaly, F. R May, and I worked on shape coexistence in the Pb region.[12,13] We found a number of cases in the $N = 105, 106, 107$ isotones and the Pb, Po, Fr, Ra, Rn isotopes. Next time I met Joe at the *International Conference on Selected Topics in Nuclear Structure*, Dubna, USSR in 1976. We discussed the ongoing work, in particular our prediction of the coexistence of spherical, oblate, and prolate shapes in ^{184,186,188,190}Pb. The calculated deformation energies are displayed in Fig. 3. The exotic phenomenon was investigated in a number of experimental and

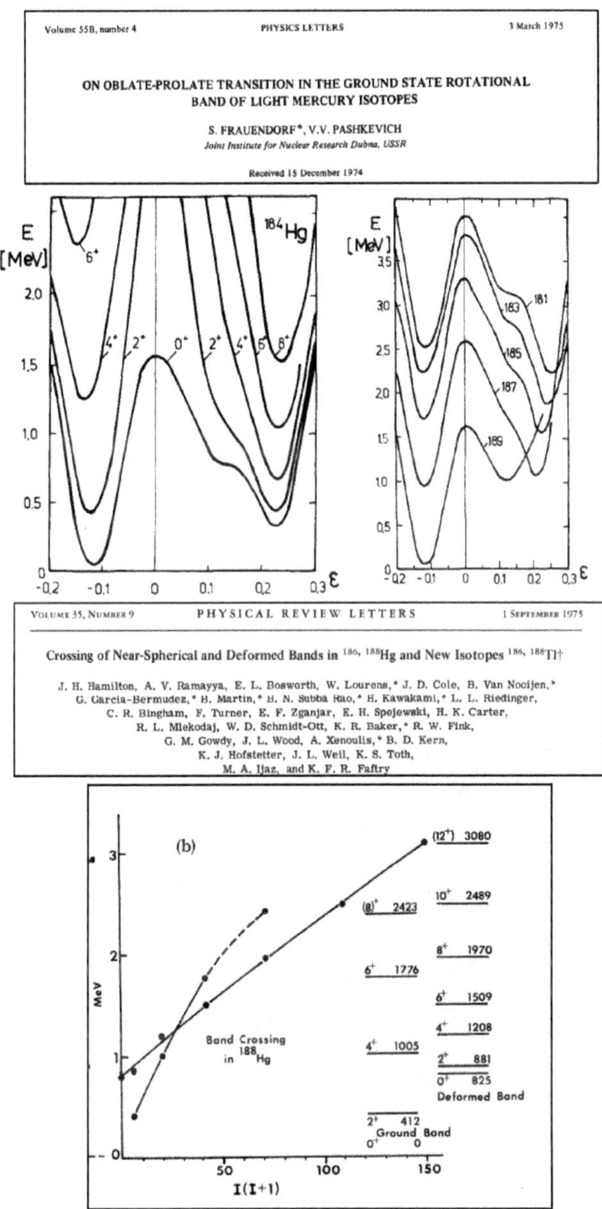

Fig. 2. Our calculated[8] (upper panel) and Joe's experimental[9] (lower panel) energies of the Hg isotopes.

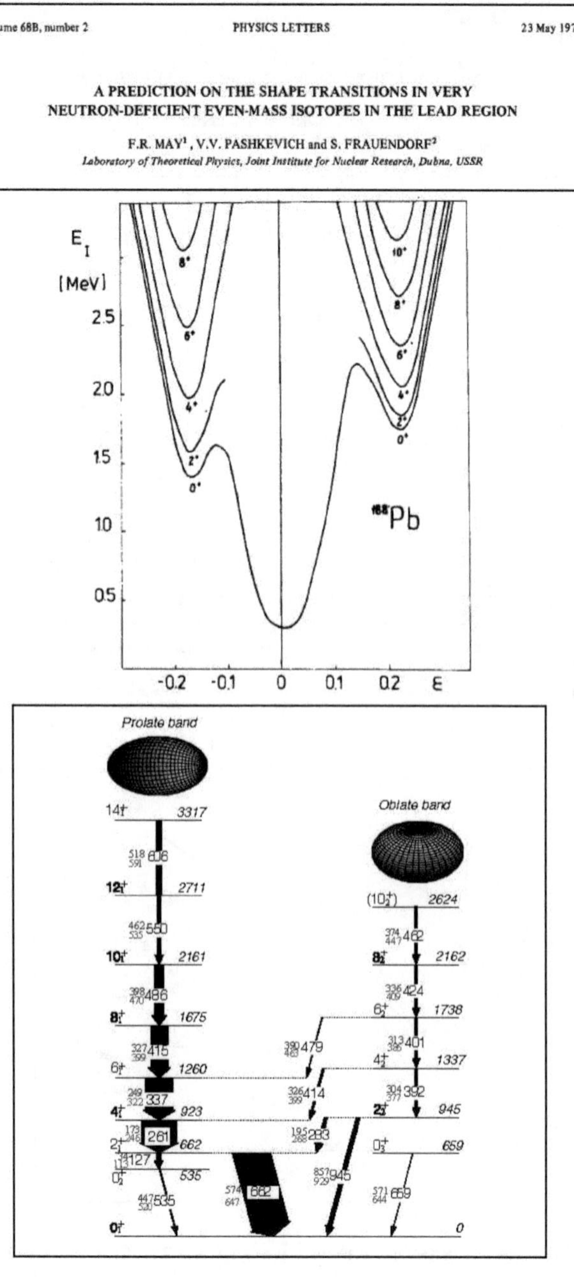

Fig. 3. Our calculated energies of ^{188}Pb as a function of the deformation parameter ε (upper panel)[12] and the most recent experimental level scheme of ^{186}Pb from Ref.[14] (lower panel). The calculated prolate minimum in ^{186}Pb lies 0.15 MeV below the oblate one.[12]

theoretical studies later on. The authors of Ref.[14] claimed the final experimental establishment of the three minima as recent as 2022. Joe kept interest in our studies and he invited me for a talk at the *International Symposium on Future Directions in Studies of Nuclei Far from Stability*, Nashville, USA, in 1979, which was a wonderful recognition of our work and a great opportunity to present our systematic investigations of shape coexistence in the Pb region (see Figs. 4 and 5). For me the conference

Fig. 4. Front row from the left: K. A. Erb, S. Frauendorf, G. N. Flerov, A. Feassler, A. H. Wapstra, J. H. Hamilton, L. K. Peker, and E. F. Zganjar.

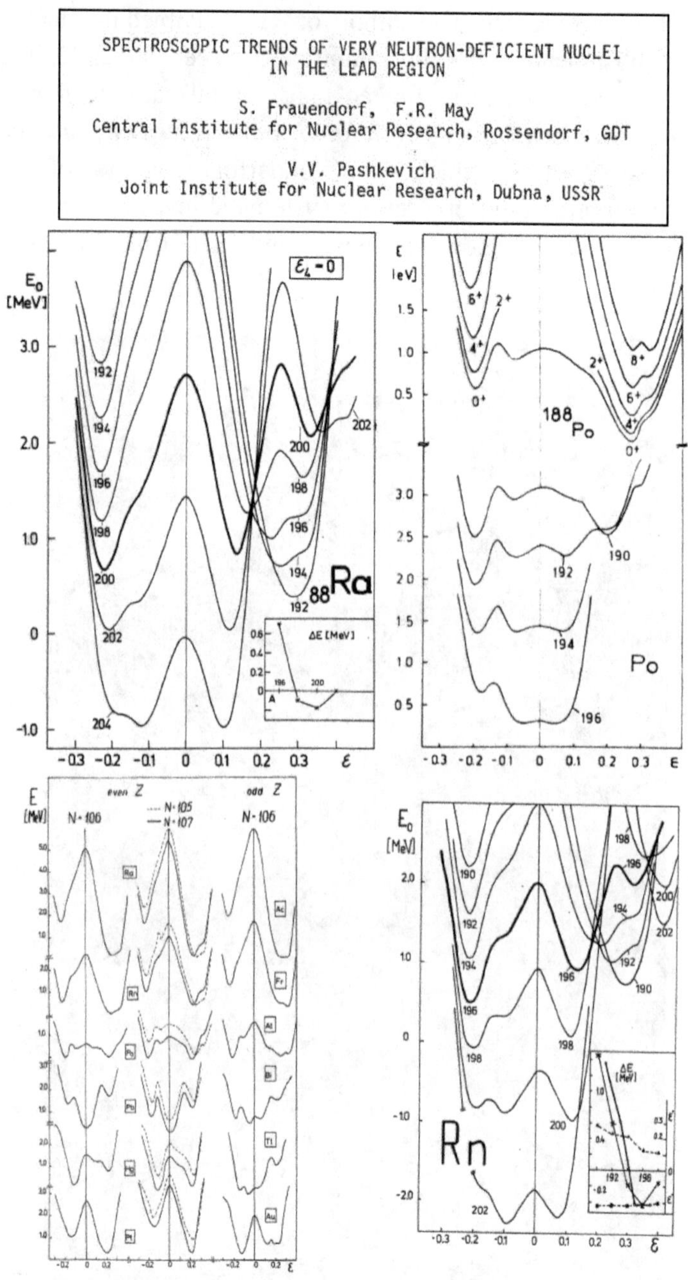

Fig. 5.   Predictions of shape coexistence in the Pb region presented at the *International Symposium on Future Directions in Studies of Nuclei Far from Stability*, Nashville, USA, in 1979. From Ref. 13

was a great inspiration and another opportunity to discuss our results with Joe, one of the world leaders exploring the field of shape coexistence experimentally.

The conference also meant stress for me. There was a group from the the JINR Dubna, headed by Prof. Georgi Flerov, one of the big Dubna shots. As many of the Russian scientists of his generation, his command of the English language was not good. When they found out that I, a Dubna guy, was there, they asked me to simultaneously translate the talks Flerov was interested in into Russian. I should have declined if this was a real option. So, I got the microphone and Flerov the headphones. It was much harder than I expected and many details got lost, however Flerov did not complain.

In the following, Joe and I met at the *International Conference on Extreme States in Nuclear Systems* (Dresden, Germany, 1980), the *XVIII Winter School on Physics* (Bielsko-Biala, Poland, 1980), *International Workshop on Nuclear Physics* (Philadelphia, PA, 1980), where we informed each other about the progress of our studies. In 1981, I had an appointment as a visiting professor at the University Tennessee Knoxville, where I continued my collaboration with Prof. Lee Riedinger's students and the ORNL group on the interpretation of their high-spin experiments. Joe invited me to present this work on a colloquium at the Physics Department of Vanderbilt University. He showed me their new results on coexistence between a weakly and strongly deformed states in the Kr isotopes. We worked out a consistent interpretation that takes the mixing of the different shapes into account[15] (see Fig. 6). Dr. R. Bengtsson then with the ORAU theory group and Dr. P. Möller and Dr. R. Nix from LANL joined the collaboration, which came up with the concept of the "Reinforcing shell gaps on the competition between spherical and highly deformed shapes":[16] Magic numbers appear for highly deformed or spherical shapes when there are both gaps in the single proton and single neutron spectra. These studies have become a standard for interpreting of shape isomerism and benchmark cases to test modern theoretical approaches.

2. Rotational Structure of Fission Fragments

Out of Joe's many activities, one is particularly appreciated by the nuclear physics community. He was the central person in organizing the series of

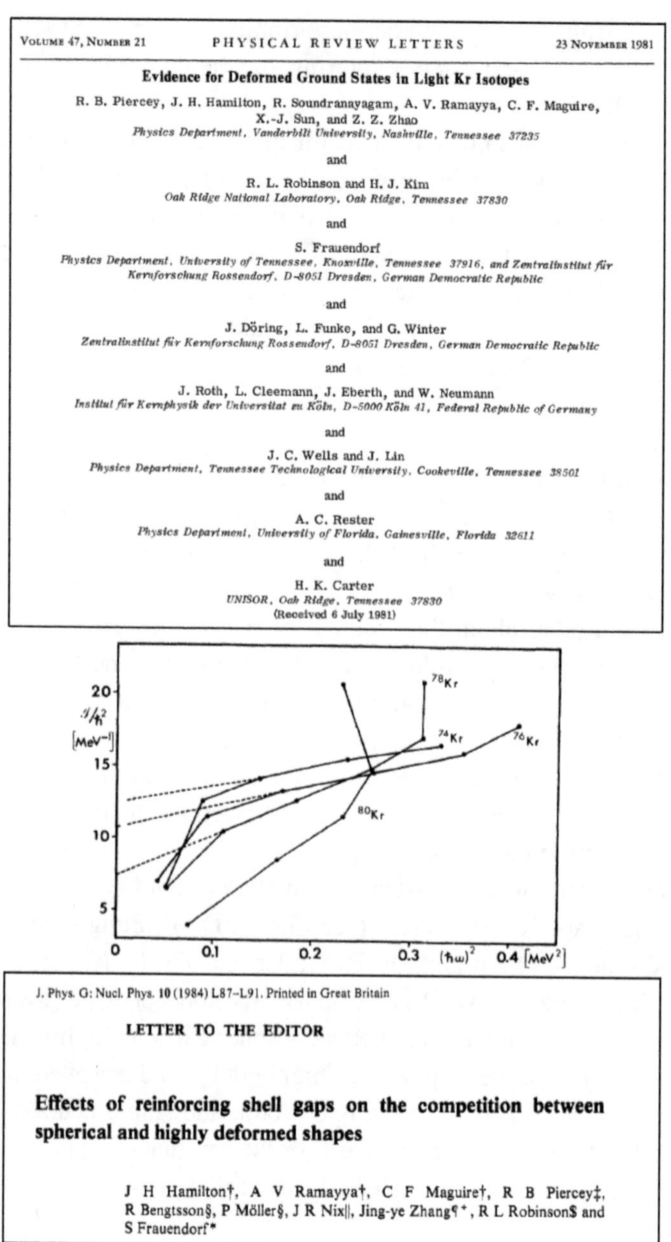

Fig. 6.   Analysis of the shape coexistence data of the Kr isotopes. Shown are the moments of inertia as functions of the square of the rotational frequency. The linear relation indicates the deformed shape at large frequency. The difference between its extrapolation and the data quantifies the admixture of the weakly deformed configuration.

the *International Conferences Fission and Properties of Neutron-Rich Nuclei* and managing to keep the financial support. The first meeting took place at the marvelous Sundial Resort on Sanibel Island to be followed by the second in St. Andrews Scotland. I guess Joe realized then that it is more fun to discuss physics under the sun of Florida than in the rain and mist of Scotland. So, the third meeting returned to the Sundial resort, which I attended in 2002 for the first time and have not missed one since

Fig. 7.   The *International Conferences Fission and Properties of Neutron-Rich Nuclei* at the Sundial Resort on Sanibel Island.

then. What a great venue to exchange ideas while enjoying Florida's beautiful nature. Thank you, Joe! Unfortunately, the seventh meeting of the series scheduled for this year had to be canceled because Hurricane Ian devastated the area. I hope that one year will be enough for restoration and the seventh meeting will take place again at the Sundial Resort in 2023.

At the meeting, I reported my work with Vesko Dimitrov on chirality of rotating triaxial nuclei (upper panels of Fig. 8). In the next break, Joe

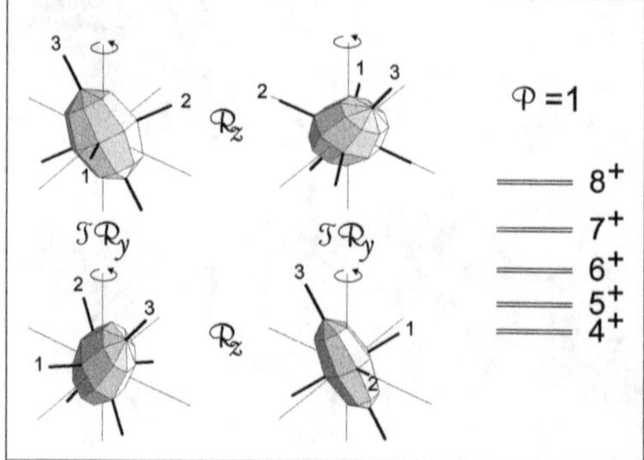

Fig. 8.   The reports of the two talks on chirality in the *Proceedings of Third International Conferences Fission and Properties of Neutron-Rich Nuclei.*[17] The upper panel illustrates the chirality of a rotating triaxial nucleus. For the left upper case, the axes 1, 2, 3 are right-handed ordered with respect to the rotational axis and for the right lower case, they are left-handed ordered. The intermediate steps in reversing the chirality are shown, where $\mathcal{R}_{y,z}$ are rotations by an angle of π about the y, z axes and $\mathcal{T}$ the time reversal operation. The left and right-handed configurations have the same energies, which is reflected by the appearance of doublets of rotational states. The lower panel shows the two rotational bands with nearly the same energy, which Joe and collaborators observed in the fission fragment ^{106}Mo and interpreted them as chiral partner bands.

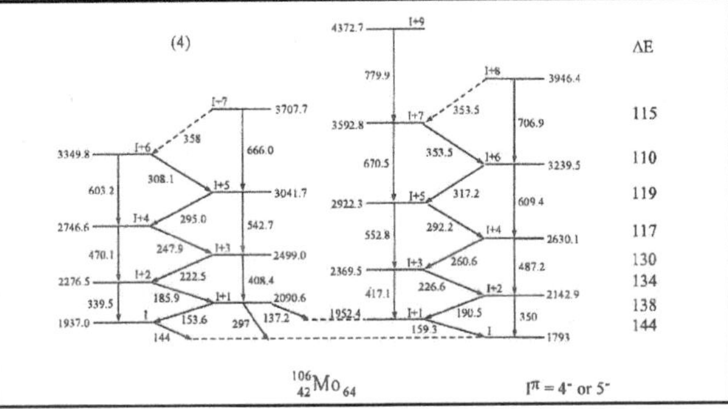

Fig. 8. (*Continued*)

approached me: "Stefan, we observed two nearly degenerate bands of negative parity in ^{106}Mo, which we interpreted as chiral partners. Zhu will give a talk on it (lower panels of Fig. 8). Let's discuss after the presentation." So far we had studied chirality in odd–odd nuclei (see upper panels of Fig. 8 and the abstract), where the angular momenta of a high-j proton aligns with the short axis, of a high-j hole with the long axis and collective with the intermediate axis. Joe's case (see lower panels of Fig. 8) was different because it concerned two rotational bands in an even–even nucleus that are built on a neutron particle–hole excitation. I was not sure if his interpretation as chiral partners was appropriate and told Joe that we will carry out Tilted Axis Cranking (TAC) calculations to check if the neutron particle–hole configuration is chiral as suggested. At this time, I had moved from Germany to the US to take a faculty position at the University Notre Dame. Dr. Vesko Dimitrov was as a visiting professor with me at Notre Dame. Our TAC calculations confirmed that the bands were chiral. They turned out to be a new type of chirality that we had not considered before. We published the findings together with the experimental results in Ref.[18] (see Fig. 9).

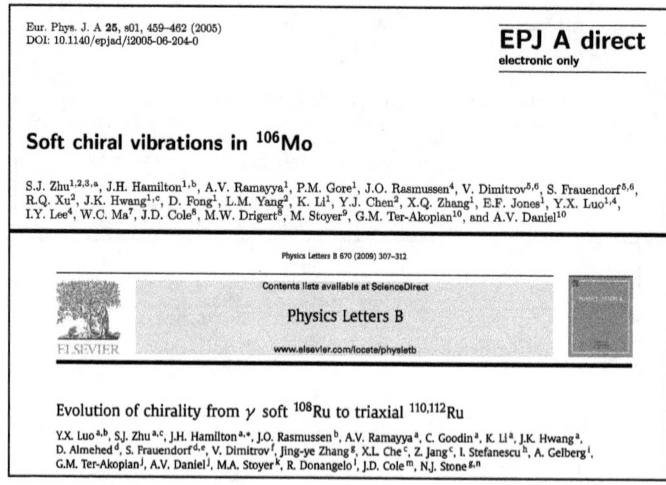

Eur. Phys. J. A 25, s01, 459–462 (2005)
DOI: 10.1140/epjad/i2005-06-204-0

EPJ A direct
electronic only

Soft chiral vibrations in [106]Mo

S.J. Zhu[1,2,3,a], J.H. Hamilton[1,b], A.V. Ramayya[1], P.M. Gore[1], J.O. Rasmussen[4], V. Dimitrov[5,6], S. Frauendorf[5,6], R.Q. Xu[2], J.K. Hwang[1,c], D. Fong[1], L.M. Yang[2], K. Li[1], Y.J. Chen[2], X.Q. Zhang[1], E.F. Jones[1], Y.X. Luo[1,4], I.Y. Lee[4], W.C. Ma[7], J.D. Cole[8], M.W. Drigert[8], M. Stoyer[9], G.M. Ter-Akopian[10], and A.V. Daniel[10]

Physics Letters B 670 (2009) 307–312

Contents lists available at ScienceDirect

Physics Letters B

www.elsevier.com/locate/physletb

ELSEVIER

Evolution of chirality from γ soft [108]Ru to triaxial [110,112]Ru

Y.X. Luo[a,b], S.J. Zhu[a,c], J.H. Hamilton[a,*], J.O. Rasmussen[b], A.V. Ramayya[a], C. Goodin[a], K. Li[a], J.K. Hwang[a], D. Almehed[d], S. Frauendorf[d,e], V. Dimitrov[f], Jing-ye Zhang[g], X.L. Che[c], Z. Jang[c], I. Stefanescu[h], A. Gelberg[i], G.M. Ter-Akopian[j], A.V. Daniel[j], M.A. Stoyer[k], R. Donangelo[l], J.D. Cole[m], N.J. Stone[g,n]

ELSEVIER

Nuclear Physics A 919 (2013) 67–98

www.elsevier.com/locate/nuclphysa

New insights into the nuclear structure in neutron-rich [112,114,115,116,117,118]Pd

Y.X. Luo[a,b], J.O. Rasmussen[b,c], J.H. Hamilton[a,*], A.V. Ramayya[a], S. Frauendorf[d,e], J.K. Hwang[a], N.J. Stone[f,g], S.J. Zhu[a,h], N.T. Brewer[a], E. Wang[a], I.Y. Lee[b], S.H. Liu[a,i], G.M. TerAkopian[j,k], A.V. Daniel[a,j,k], Yu.Ts. Oganessian[j], M.A. Stoyer[l], R. Donangelo[m], W.C. Ma[n], J.D. Cole[o], Yue Shi[p], F.R. Xu[p]

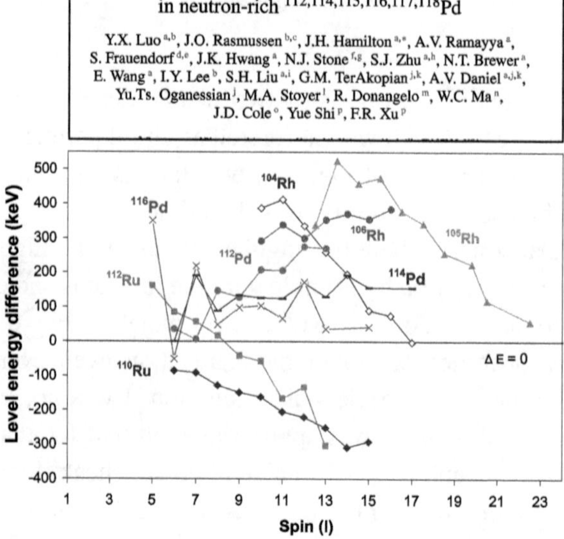

Fig. 9. Selection of the joint publications.[18–20] The lowest panel compares the energy distance between chiral partners suggested in our studies of the Pd isotopes[20] and Ru isotopes[19] as well as of the Rh isotopes from the literature.

This was the beginning of an extremely fruitful collaboration on studying the structure of neutron-rich nuclei, which has continued until now. On the experimental side, Joe organized an extended collaboration which analyzed the γ spectra emitted by the fragments of spontaneous fission of ^{252}Cf by means of the GAMMASPHERE multidetector array. My contribution was the theoretical analysis of the structure of these neutron-rich nuclei in the mass 100 and 130 regions. The topics comprised wobbling and chirality as manifestations of a triaxial shape, deviation from reflection symmetric shapes caused by the condensation of octupole phonons and the quasiparticle alignments caused by rotation. The results were published as six articles in well-recognized journals[18–23] (see Fig. 9). Joe was enthusiastic in deseminating the results of our work at international conferences, which is reflected by sixteen joint contributions to conference proceedings.

Joe and I discussed our projects at several mutual visits at Vanderbilt and Notre Dame as well as at the meetings at the Sundial resort. Otherwise, our discussions were telephone calls, the early form of nowadays e-communication. Often, the phone on my desk rang with Joe on the line or it blinked announcing Joe's voice message: "Stefan, please give me a call." When I called Joe, he was usually busy with one of his many obligations. However, his secretary was very efficient in arranging a conversation in the short term.

Joe, working with you has been an extraordinary experience and source of inspiration. Happy retirement and thank very much!

References

1. V. M. Strutinsky, Shell Effects in Nuclear Masses and Deformation Energies, *Nucl. Phys. A* **95**, 420–442 (1967).
2. V. M. Strutinsky, "Shells" in Deformed Nuclei, *Nucl. Phys. A* **122**, 1–33 (1968).
3. J. Bonn *et al.*, Optical Pumping of Neutron Deficient ^{187}Hg, *Phys. Lett.* **36B**, 41–43 (1971).
4. J. Bonn *et al.*, Sudden Change in the Nuclear Charge Distribution of Very Ligh Mercury Isotopes, *Phys. Lett.* **38B**, 308–311 (1972).
5. W. E. Collins *et al.*, Levels in ^{72}Se Populated by ^{72}Br, *Phys. Rev. C* **9**, 1060–1063 (1974).
6. J. H. Hamilton *et al.*, Evidence for Coexistence of Spherical and Deformed Shapes in ^{72}Se, *Phys. Rev. Lett.* **32**, 239–243 (1974).
7. J. H. Hamilton *et al.*, Lifetime Measurements to Test the Coexistence of Spherical and Deformed Shapes in ^{72}Se, *Phys. Rev. Lett.* **36**, 340–342 (1976).
8. S. Frauendorf, V. V. Pashkevich, On Oblate-Prolate Transition in the Ground State Rotational Band of Light Mercury Isotopes, *Phys. Lett.* **55B**, 365–368 (1975).
9. J. H. Hamilton *et al.*, Crossing of Near Spherical and Deformed Bands in ^{186,188}Hg and ^{186,188}Tl, *Phys. Rev. Lett.* **35**, 562–565 (1975).
10. N. Rud *et al.*, Lifetimes in the Ground-State Band of ^{184}Hg, *Phys. Rev. Lett.* **31**, 1421–1423 (1973).
11. D. Proetel *et al.*, Evidence for Strongly Deformed Shapes in ^{186}Hg, *Phys. Rev. Lett.* **31**, 896–898 (1973).
12. F. R. May, V. V. Pashkevich, and S. Frauendorf, A Prediction on the Shape Transitions in Very Neutron-Deficientevenmassisotopes in the Lead Region, *Phys. Lett.* **68B**, 113–116 (1977).
13. S. Frauendorf, F. R. May, and V. V. Pashkevich, Spectroscopic Trends of Very Neutron-Deficient Nuclei in the Laed Region, *Proceedings of the International Symposium on Future Directions in Studies of Nuclei Far from Stability,* Nashville Tennessee, September 10–13, 1979, eds. J. H. Hamilton, E. H. Spejewski, and C. R. Bingham, North-Holland Pub. Co., pp. 133–149 (1980).
14. J. Ojala *et al.*, Reassigning the shapes of the 0^+ states in the ^{186}Pb nucleus, *Commun. Phys.* **5**, 213–222 (2022), https://doi.org/10.1038/s42005-022-00990-4.

15. B. B. Piercey *et al.*, Evidence for Deformed Ground States in Light Kr Isotopes, *Phys. Rev. Lett.* **47**, 1514–1517 (1981).

16. J. H. Hamilton *et al.*, Effects of Reinforcing Shell Gaps on the Competition Between Spherical and Highly Deformed Shapes, *J. Phys. G: Nucl. Phys.* **10**, L87–L91(1984).

17. *Proceedings of theThird International Conferences Fission and Properties of Neutron-Rich Nuclei, Sanibel Island*, eds. J. H. Hamilton, A. V. Ramyya, and H. J. Carter, World Sci. Pub. Co. (2003), pp. 93–100, ibid. pp. 191–206.

18. S. J. Zhu, Soft Chiral Vibrations in 106Mo, *Eur. Phys. J. A* **24**, 459–462 (2005).

19. Y. X. Luo *et al.*, Evolution of Chirality from γ-soft ^{108}Ru to Triaxial ^{110,112}Ru, *Phys. Lett. B* **670**, 307–312 (2009).

20. Y. X. Luo *et al.*, New Insights into the Nuclear Structure in Neutron-rich ^{112,114,115,116,117,118}Pd, *Nucl. Phys. A* **919**, 67–98 (2013).

21. X. Y. Luo *et al.*, First Observation of a Rotational Band and the Role of the Proton Intruder Orbital $\pi 1/2^+[431]$ in Very Neutron-Rich Odd-Odd ^{106}Nb, *Phys. Rev. C* **89**, 044326-1–13 (2014).

22. X. Y. Luo *et al.*, First Observation of Collective Rotational Bands in Neutron-Rich ^{142}La and the Study of Octupole/Triaxial Deformations in (142)'La-143, *Int. J. Mod. Phys. E* **30**, 2150037 (2021).

23. B. M. Musangu *et al.*, Chiral Vibrations and Collective Bands in ^{104,106}Mo, *Phys. Rev. C* **104**, 064318-1–18 (2021).

Joe Hamilton and the Little I

Da Hsuan Feng

A photo taken by one of Joe's first graduate students Lee Riedinger when we all were at a conference in China's Institute of Atomic Energy. The name of the conference was *Physics at the Tandem*! I think it was some four decades ago in the 80s. Obviously, the photo was taken when I was still in my youth. The irony is that even today, Joe still looks exactly like that, in his eternal youth look! I showed this photo to a friend in China recently, and his observation was that "the bicycles in the background all looked brand-new. In the 80s bicycles were absolute transportation necessities and were very expensive household items (a full professor salary then was 80 RMB per month). So, most bicycles were used a long time. The fact they looked new here must mean that having distinguished foreigners visiting, the hosts wanted to put on the best front and to make sure that only brand-new bicycles were on display!"

Sometime in the 90s, Joe and I were traveling together in Spain (to or from) a conference. I looked older and Joe looked identical to the earlier picture.

A recent photo of Joe. I rest my case.

As one of the most distinguished physicists/teachers, and hitting the age of 91 and having been at Vanderbilt University for 65 years, Joe is an unsinkable individual.

My friendship with Joe began around 1979, some forty-three years ago. I was a young assistant professor who had the pleasure and honor of collaborating with a renowned senior nuclear theorist in the country, Tom Pinkston of Vanderbilt University. Tom was the co-author of the famous paper which led to the so-called Pinkston–Satchler equation which was utilized in calculating light nucleon integrals! When my colleague Michel Vallieres and I at Drexel University developed a tri-diagonalization method of calculating the two nucleon overlap functions, I pushed up my courage and sent a copy to Tom. To my amazement, not only Tom responded (enthusiastically, I might add), but he even suggested that we should collaborate on this project. This led us to nearly several decades of collaboration, until sadly his death in 2001!

Once I started to work with Tom, naturally I had the great pleasure of visiting Vanderbilt. I remember in one of my visits, maybe even it was the first one, Tom introduced me to Joe. I quickly realize that Joe is 13 years older than I am. He is what we Chinese would refer to endearingly as an "Elder Generation" or "长辈". Knowing his scientific accomplishments even then, I was in awe of him. While I was pretty sure Joe did not understand Chinese then, the way Joe treated me during our first meeting, and ever since, it was and still is abundantly clear to me that he understands the responsibility of being a 长辈 well! Being a teacher is indeed second nature to him (maybe even first nature), as I am sure Lee Riedinger, being one of Joe's earliest doctorate students, who is also a highly accomplished physicist and administrator himself, could and will testify to that!

Throughout our many decades of collaboration, Joe and I had numerous conversations. It should be underscored that while he and I were both good friends and we are both physicists, I have noticed that whenever Joe mentioned about our mutual acquaintances, he would always define his friendship with that person as to how many papers they have published together. Well, in this sense, our friendship is quite different, maybe even unique! This is because Joe and I, as far as I can remember in his well over 1,000 publications, and we have had many in-depth scientific discussions together, he and I have NEVER co-authored a paper together, except co-editing a conference proceeding. I am proud to tell the world that our friendship is best characterized as what the ancient writings of Zhuang-Zi (庄子), namely 君子之交淡如水, or the "friendship between gentlemen is pure as clear water!"

Even if one were to only make a cursory observation of Joe's life, you could easily come to the realization that Joe is deep down in his psyche, a man of multidimensions and of passion for humanity, something which is abundantly written by friends and colleagues of Joe with great passion from all corners of the world!

I have always professed that one of my ultimate aims in my life is that whomever I meet, I hope I am able to change the life trajectory of that person, albeit minute. I think Joe certainly is able to do that, except more. He not only will change the life trajectory of the person he meets, but he also changes it significantly, for the better!

I am deeply honored to have the privilege of knowing someone who is as grand and as magnanimous as Joe!

Joseph H. Hamilton: A Grand Master of Nuclear Science

Mark A. Riley

Department of Physics and The Graduate School,
Florida State University,
Tallahassee, FL 32306, USA
mriley@fsu.edu

Few nuclear scientists have had as remarkable an impact on the field as Joseph H. Hamilton. His presence at seemingly every significant national and international meeting, and always with interesting new results, through the decades, is staggering. His limitless enthusiasm for discovery has inspired generations of experimentalists and his discoveries have challenged and inspired theorists for more than half a century. But perhaps Joe's greatest legacy will be his incredible humanity, his unbelievable mentorship skills, and his absolute belief in the importance of collaboration without borders. He is truly a Grand Master of Nuclear Science. Joe has had a major impact throughout my career and in the following I reflect on a few of our fun and memorable interactions throughout the years.

1. Introduction

Let me begin with the introductory paragraph I wrote in a support letter a decade ago for Joe for one of the highest scientific awards in the US. "It is with very great pleasure that I support the nomination of Dr. Joseph

Hamilton..... To say he fulfills the stated criteria would be an understatement. I know of no other scientist who has the breadth, depth and sustained impact of performing world-class science, and through whose management astuteness has brought enormous benefits at DOE facilities enabling many hundreds of other scientists to also perform groundbreaking work. That he has continued this level of productivity over seven decades is simply incredible. He is a remarkable and unique individual whose achievements are fully in line with the expectations of this most prestigious award." Since the award criteria overlapped superbly with Joe's career trajectory stating that the nominee must have "contributions and achievements that (i) must show significant innovation and discovery; (ii) must be identifiable as lasting, impactful, and substantial; and (iii) must demonstrate prominent scientific, technical, management or policy leadership", we felt Joe would be a "shoe-in" to win this award. Unfortunately, it was not to be, but by goodness, he must have come close. Still, Joe was not perturbed at all and shrugged it all off in his usual inimitable way. He had won so many other major national awards already and commented in his laid-back southern drawl that "there are some amazing scientists out there you know!"

I will not quote further from my detailed five-page letter where I summarized many of Joe's amazing scientific accomplishments since they are repeated elsewhere in this volume by those who were there "in the room" with him as historic events unfolded in real time. I have known Joe for five decades and collaborated with him on a number of projects. He has had a major impact on my career development and so I thought I'd reflect on a few of our fun and memorable interactions through the years.

2. Interactions through the Decades

As a young graduate student at Liverpool University working with the John Sharpey-Schafer and Peter Twin group in the early 80s, in gamma-ray spectroscopy and nuclear structure, I had obviously heard of Professor Hamilton and his ground-breaking investigations into nuclear shape co-existence. Our studies centered on shape co-existence too but at much higher spins, where predictions of superdeformation and dramatic angular momentum-induced shape transitions had been made by the Lund and

Warsaw groups. These were exciting times. The brand new Daresbury Laboratory accelerator had just come online, and we were also involved, with the Niels Bohr Institute and Manchester University groups, in the development of a new form of high-resolution multidetector system, in which the Ge(Li) detectors were surrounded by a large scintillator detector in an escape suppression mode. These early TESSA (The Escape Suppresssed Spectromter Array or later called Total Energy Summed Spectrometer Array) arrays helped revolutionize the field with their far superior "peak-to-background" data collection ratio and soon after every major laboratory had put together a "TESSA-like" system. At Oak Ridge National Laboratory, Joe was a leader in incorporating escape or Compton-suppressed Ge detectors into the famous Spin Spectrometer designed by Demetrios Sarantities. I remember the first time I heard Joe give a presentation was him showing new data from this system.

Lee Riedinger from UT Knoxville, a great friend and collaborator of Joe's, who was also one of the major drivers of the Compton-suppressed Ge detectors and the Spin Spectrometer array at ORNL, spent a summer with the Liverpool group in 1983 while I was a graduate student and I got to know him quite well since my PhD thesis centered on the $N = 90$ nuclei — ^{156}Dy and ^{158}Er, which were isotones with Lee's favorite ^{160}Yb nucleus from his NBI work. We even collaborated on an experiment at Daresbury with TESSA2 on ^{162}Hf and published a paper on these results.

Lee had also spent an enormously productive sabbatical at the NBI in 1979/80 and was also a close collaborator and friend with Jerry Garrett at the NBI who I worked most closely with during my post-doc years there in 1985 and 1986. It was through these connections with Lee Riedinger and Jerry Garrett that I was offered a post-doc working at ORNL with the Compton Suppressed Ge Spin Spectrometer array from 1987 to 1988.

At ORNL, experiments using the abovementioned system were taking place at a backbending pace, many of which involved the Joe Hamilton and Ramayya group. Joe was a prime mover in all things at the Holifield Laboratory and I had a great office of my own in the fancy Joint Institute for Heavy Ion Research building. It was a most exciting period with a flowering of new physics results, and I got to attend Joe's many scientific presentations at the JIHIR during this period. His enthusiasm was compelling and contagious! I also remember Jerry Garrett arriving towards the end of my stay and being present for some of the early meetings about the transformational upgrade to the Holifield Rare Isotope Beam Facility (HRIBF) in which Joe and Jerry were leading lights. I left ORNL at the end of 1988 to take up a faculty position at Liverpool University.

My next major interaction with Joe was at the International Nuclear Structure Conference in 1990 held in Oak Ridge. This was a huge meeting, with Joe (and Lee) being members of the main Organizing Committee. It was rather significant for me in that it dramatically changed my life, and that of my family, as I will attempt to illustrate in the following. The meeting was buzzing with a dizzy spectrum of new results related to a variety of nuclear shapes, especially high-spin superdeformation. I remember Joe giving a presentation entitled "New Insights into Nuclear Structure from Nuclei Far from Stability" and he included data from a recent ORNL CSS-Spin Spectrometer experiment searching for superdeformation in ^{186}Hg. An interesting sideband had been observed with a higher-than-usual moment of inertia and Joe conjected it may be superdeformed. I had presented a talk about superdeformation in ^{193,194}Hg early and so I was ultra-interested in what Joe had to say. I stood up in question time and posed to him an alternative interpretation of his data since I was not fully convinced of his own interpretation. We had a friendly exchange and chatted later in the coffee break along with Jerry Garrett. Jerry explained that we were both really saying the same thing but using slightly different terms, LOL! It was exactly during this coffee break discussion with Joe that Jerry spotted Sam Tabor in the crowd and Jerry politely asked Joe if we could talk to Sam since he knew of a tenure track faculty position opening at some strange place called Florida State University which may be of interest to me. Joe bid us farewell and we raced off to catch Sam. I had seen Sam give presentations at the NBI May meetings and other conferences but I had never spoken to him before and knew nothing about FSU. But I am sure Sam seeing me talking with Joe and then being introduced by Jerry was highly influential.

I managed to obtain that position in 1991 and have been at FSU ever since. With Sam Tabor and Jurg Saladin from Pittsburgh, we built the largest Compton-Suppressed Ge array at any university laboratory in the world at that time. Throughout the 1990s, Joe and Ramayya (and students) would often come down to Tallahassee to do experiments. With no "official" Program Advisory Committee, Sam and I were always happy to see our Vanderbilt friends and collaborate on some new physics. It was brilliant to see Joe and Ramayya sitting in the control room surrounded by students and online printouts of the data we were collecting. This is one of the most fun aspects of science.

As a young assistant professor, with FSU being a member of Oak Ridge Associated Universities (ORAU), I applied for and received an ORAU Junior Faculty Enhancement Award in 1993. This played a major part in my promotion to associate professor with tenure soon after. I remember proudly telling Joe since I suspected he may have had a hand in it and him saying nonchalantly "Yes, I really enjoyed reviewing that proposal, it was quite good you know".

As I said, Joe's group would often come down to Tallahassee, and my group would often go up to ORNL and perform science there, spending many nights at the "Hamilton Hilton". However, one particular stay there will forever be etched into my memory. I was part of a major Nuclear Science Advisory Review Committee touring each of the DOE nuclear structure labs and we were having breakfast at the JIHIR when the terrible events of September 11, 2001, unfolded on the TV in front of us! Our meetings were canceled, the lab closed and locked down, and we all did our best to find a way to get back home. I somehow managed to get one of the very last rental cars available just outside Oak Ridge and it was below $500!

In the second half of the 90s and into the next decade, the new generation of escape-suppressed Ge arrays came online with Gammasphere in the USA and Euroball in Europe. I was involved in performing experiments on Gammasphere and Joe was too, of course, most famously with his ground-breaking Cf fission studies of neutron-rich nuclei. We were all so jealous of Joe's brilliant idea to perform these measurements, not only because the results were spectacular but also because he didn't have to fight for beamtime in the same way since he didn't need a beam! They could be run when other non-Gammasphere runs were taking place and even during the accelerator "upgrade or maintenance" periods. I remember meeting after meeting where Joe unveiled new result after new result, bringing light and understanding to these previously uncharted waters. This physics story is covered elsewhere but I do have a little contribution to tell of which I am quite proud.

In 2011, Joe invited me to give a colloquium, about recent results from Gammasphere, to his department at Vanderbilt. He and I are both "car people" and I was mega-impressed when he picked me up from the airport in his purple Plymouth Prowler "Batmobile". Perhaps even more impressive was his number plate "117Tn", see Fig. 1. We spent a marvelous day discussing physics and the analysis of Gammasphere data. Since my studies were at ultra-high spin and involved high-multiplicity events, we had been using the latest Radware software package to build four-fold

Fig. 1. Collage of photos of JHH and me through the years, plus his famous purple Plymouth Prowler "Batmobile" and number plate. The photo of JHH and Lee Riedinger is significant since I remember catching Lee out much to Lee's dismay and Joe's delight. The photo was taken before the game when Lee was still smiling ☺! The 2021 photo, which includes FSU's Kirby Kemper, is from Joe's visit to a SESAPS meeting in Tallahassee.

coincidence "hyper-cubes". I demonstrated to Joe and his group, who had been analyzing three-fold coincident cubes, that being able to set that "one extra coincidence gate", even with the loss of statistics, could have a dramatic effect in removing ambiguities and cleaning things up. Joe realized the potential immediately. He had collected quite enormous data sets in terms of statistics, and although the multiplicity was lower than my in-beam studies, he had his post-doc set to work straight away to explore this new possibility. Sure enough, he found that building hypercubes from his fission data could be really useful too, and as always the gracious gentleman, he remembered and thanked me for convincing him to try it!

Joe, as we all know, is one of the nicest people you can ever interact with, however, I did on one occasion perhaps "annoy" him a little. It was in November 2016 when the International Union of Pure and Applied Chemistry (IUPAC) approved the name tennessine for element 117, as well as the naming of another three elements 113, 115, and 118. The morning when the BIG official IUPAC announcement was on all the newspaper front pages and on the morning TV shows, I wanted to congratulate Joe and called him up sharply at 8:00 am since I knew he was always an early riser. Unfortunately, I had forgotten about the time difference between Tallahassee and Nashville and I think he was just having his breakfast. Still, he thanked me graciously and we had a brief chat but I suspect he would have preferred a call later in the day!

It was around this time, that Joe and I collaborated on something other than nuclear structure physics. We were both members of a national search committee to find the next Editor-in-Chief for the American Physical Society journal enterprise. The stature and significance of these journals in the diffusion of knowledge to the world makes the APS EiC position one of the most important roles in all of physics. We carefully studied the applications and discussed in great detail as a committee who we might invite for interviews. It was an intense and extremely time-consuming process but I will never forget how Joe's superb insights played such a significant part in our deliberations and eventual selection.

Finally, Joe recently, post COVID, came down to Tallahassee as an invited guest speaker at the Southeastern APS meeting at FSU in November 2021. He gave a typical brilliant presentation entitled *60 Years of Nuclear Physics in the South-East*. His lecture was extremely well attended and he was in fine form. His retrieval of details through the decades was simply remarkable. FSU's Kirby Kemper remarked at the end of Joe's talk *We have just heard from a living legend.* At the conference dinner, the great story-telling continued and it was an absolute joy to sit and listen to Joe and Kirby exchange tales of historic events and people!

Joe, you are truly a nuclear science Grand Master and an amazing inspiration to us all! THANK YOU MY FRIEND! ☺

https://doi.org/10.1142/9789811296963_0008

The Discovery of the Element 117 and Naming it Tennessine

J. H. Hamilton* and Yuri Oganessian[†]

*Department Of Physics And Astronomy, Vanderbilt University, Nashville, TN 37235, USA

[†]Joint Institute for Nuclear Reactions, Dubna, Russia

A brief history of the formation of the collaboration of Vanderbilt, the Flerov Laboratory in Russia, and Oak Ridge National Laboratory and then Lawrence Livermore National Laboratory joining to obtain a target of ^{249}Bk to be used to discover the new element $Z = 117$ is described first. The production and separation of the ^{249}Bk then follow. Two experiments and their results from the bombardment of ^{249}Bk with ^{48}Ca in the Dubna cyclotron and new results of confirmation in GSI in Germany are presented. Finally, the process of the certification of our discovery of the new element 117 and naming it for the State of Tennessee are described.

In the early 1960s, the prevailing model for nuclei was the liquid drop model. This model predicted that nuclei (elements) beyond atomic number around 102 could not exist because they would spontaneously fission with lifetimes of 10^{-14} s. Then, in the middle 1960s, a Russian theorist calculated that there are new magic numbers for protons and neutrons to form deformed double magic numbers for 102,152 and 108,162 and spherical double magic nuclei for Z, N with $Z = 114$–126 and $N = 184$. The lifetimes for the spherical double magic nuclei could be very long to form an "Island of Stability". These predictions were quickly supported by calculations by German and Polish theorists. See Fig. 1.

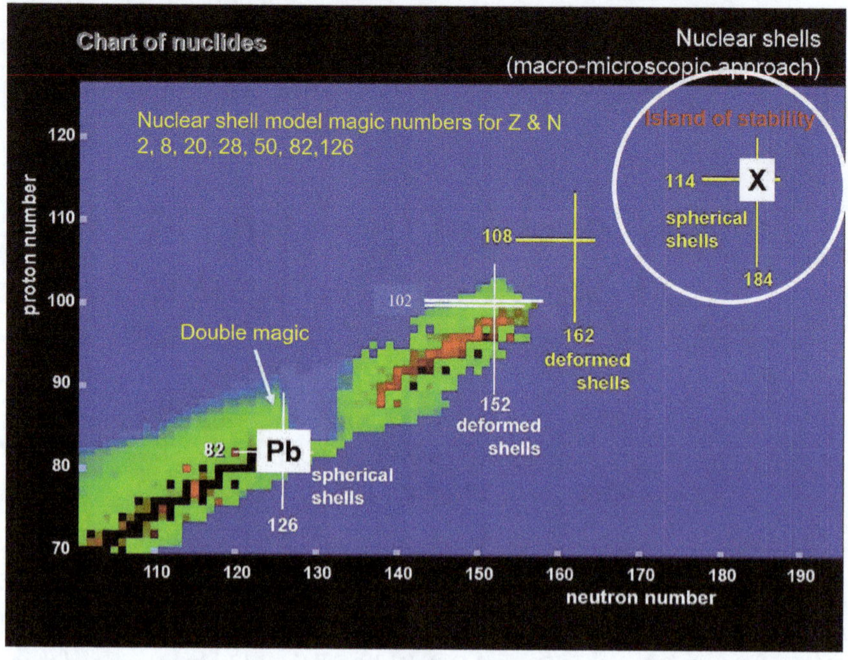

Fig. 1.

Immediately, nuclear laboratories in the US, Germany, and Russia began searches to identify these new elements. At GSI in Germany, they bombarded ^{208}Pb ($Z = 82$) and ^{209}Bi ($Z = 83$) with ^{40}Ca ($Z = 20$) to Zn ($Z = 30$) to discover the new elements 107 to 112 to confirm the deformed magic numbers. These new elements only had N up to 165, far removed from the 184 in the predicted island of stability. Their half-lives were increasingly short, down to 0.7 ms for 112.

The Russian laboratory in Dubna which had a dedicated cyclotron got Lawrence Livermore National Laboratory to join them and to bring long half-life actinide targets from Np ($Z = 93$) to Cf ($Z = 98$) to bombard them with highly enriched neutron-rich ^{48}Ca prepared in Russia to produce a few atoms of 113, 114, 115, 116, and 118. The International Union of Pure and Applied Chemistry and Physics accepted the discoveries of 114 and 116 where they had good numbers of atoms but not 113, 115, and 118 where they had only 3 atoms each. There was no long-lived target of Bk isotopes to make 117, the longest lived ^{248}Bk had a half-life of only 320 days and was difficult to make.

The Russians were unsuccessful in getting ^{249}Bk from Oak Ridge National Laboratory, where it could be made in their High Flux reactor. Having spent nearly 15 years doing nuclear structure research with Hamilton and knowing of all the projects Hamilton had done at ORNL, Oganessian approached Hamilton to join them and to seek to obtain a ^{249}Bk target from ORNL to discover the 117th element.

Together we went to ORNL in May 2005 to talk with the High Flux Reactor scientists about getting the target. We were told they could make such a target anytime for about three million dollars. But then they said from time to time commercial companies order a source of ^{252}Cf and ^{249}Bk is a byproduct of making the Cf. For only 600,000 dollars, they could chemically separate the Bk for us. Unfortunately, they had no order for such. For the next three years, Hamilton called every three months the ORNL scientists to see if an order had been placed. Finally, in August 2008, the answer was yes and that the target would come out of the reactor in December 2008. Fortunately, in September of 2008, Hamilton's Department Chair had organized a Symposium to honor Hamilton's 50 years of research and teaching at Vanderbilt. Academician Oganessian came to talk about their research together and Jim Roberto, then Deputy Director of ORNL, came to talk about how Joe's projects there had transformed ORNL.

Hamilton arranged a luncheon for Oganessian to describe to Roberto how he and Joe would like ORNL to form a collaboration with them to get a target of ^{249}Bk and discover the important new element 117. Dr. Roberto was very excited by this prospect. He asked Joe to get one of his nuclear structure collaborators at ORNL to join him and write a proposal together for $500,000 from a special program at ORNL for new projects. They did and it was awarded. To obtain the other needed $100,000, Hamilton called another of his research collaborators at LLNL and asked if they could provide the $100,000 and join the project which they did.

The target came out of the reactor in December. A waiting period of three months was needed to let the short-lived radio activities decay away so the material was safe to handle in hot cells. We both went to ORNL to see the target in a long aluminum tube come into the first hot cell. See us looking at the tube behind the six feet of glass shielding in the first of six hot cells where the chemical separations would take place. See Fig. 2. After another three months of chemical separations in the six cells, 22 milligrams of ^{249}Bk were obtained. See Fig. 3.

This material was encapsulated in a lead container for shielding and taken by Delta Airlines to Moscow. With no papers to identify the material, it was sent back to the US. A second time across Delta brought paperwork but it was not sufficient and it was sent back again. The next time the paperwork was accepted and the material was sent to a reactor institute with hot cells to prepare the target material on a six-sector wheel for the cyclotron. Shown in Fig. 4. To keep the material from being evaporated by the large energy brought in by the ^{48}Ca, the wheel was rotated 1,700 times per minute and the beam was wiggled up and down on the target as well. The neutron-rich ^{48}Ca is only 0.7% abundant in nature and one needs a nearly pure beam. The only place in the world where stable isotopes are currently separated to give the very costly, highly enriched ^{48}Ca is in Russia.

Fig. 2.

Fig. 3.

position sensitive
strip detectors
"veto" detectors
TOF-detectors
SH
recoil
side
detectors
recoils
48Ca-ions
gas-filled
chamber
detector
station
22.5°
rotating
entrance
window

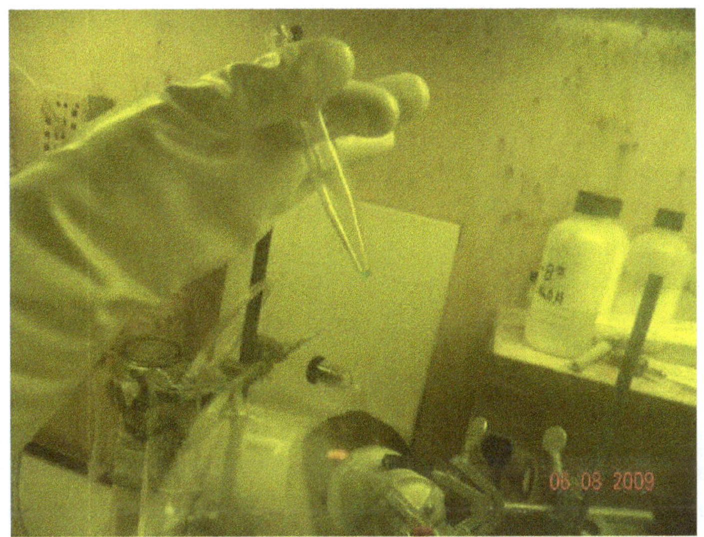

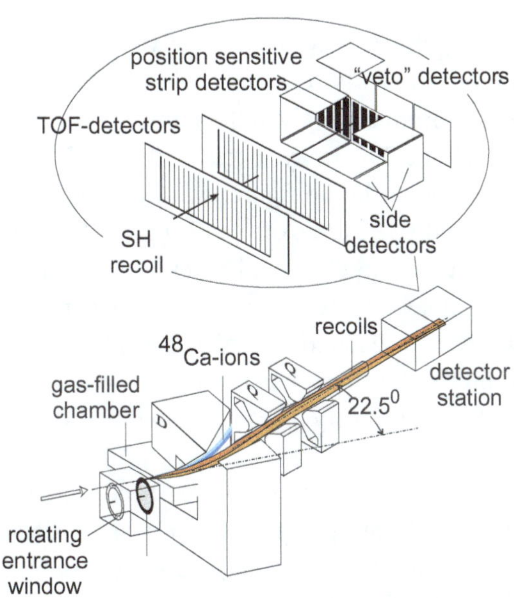

Fig. 4.

A cyclotron in Dubna was used to accelerate the ^{48}Ca to sufficient energy to overcome the repulsive Coulomb force with the BK to make element 117 in a fusion reaction ^{48}Ca $(Z = 20)$ + ^{249}Bk$(Z = 97)$ → $117 + x$ neutrons. The detector box is also shown in Fig. 4 and consists of a series of detector pixels 1 mm × 1 mm. After the reaction, the products are passed through a gas-filled magnetic separator to isolate the 117 atoms from the unwanted other products. When the 117 recoil nucleus hits a pixel, a signal for the energy deposited is generated. If the recoil is a real 117 nucleus, it will be followed by a series of alpha decays (He $Z = 2$) and ending with spontaneous fission. Since there are scattered alpha particles, all the decay alphas and at least one of the fission products must be in the same pixel with 117 or if the alpha scatters out and is caught in a side or top detector in proper time sequence.

On the last of July and the first of August, Professors Ramayya and Hamilton went to Dubna for the start of the experiment and Hamilton went again in September. Whenever a possible event was observed, it was also sent by computer to ORNL, VU, and LLNL for analysis. All groups needed to concur that an event was a real 117 event. The target had a few times 10^{19} atoms of ^{249}Bk and these were bombarded with a few times 10^{19} Ca atoms. The first published events are shown in Fig. 5. The number of observed 117 events shown is consistent with the expected cross section (probability) for the production given the number of atoms of Bk and Ca. Note in the figure all the 294 decays occur within 19.3 +/– 0.3 mm. In this first experiment as shown in the figure, we observed one atom of ^{294}117 with 177 neutrons and five events of ^{293}117 with 176 neutrons.

Our results were sent to Physical Review Letters on a Friday and we were called on Monday saying our paper was accepted for publication (PRL generally takes 1–2 months to give acceptance!). The paper was selected as an Editor's choice to be read by everyone. *The New York Times* broke the news of our discovery and within two weeks over 250 newspapers and TV stations worldwide had carried our discovery!!

For a new element to be confirmed, there needs to be some independent verification. The Department of Energy was so delighted with our discovery that they gave us free of charge a second sample of ^{248}Bk with the condition that half of the sample be given to GSI to enable them to confirm our discovery. We repeated our experiment and obtained 3 more

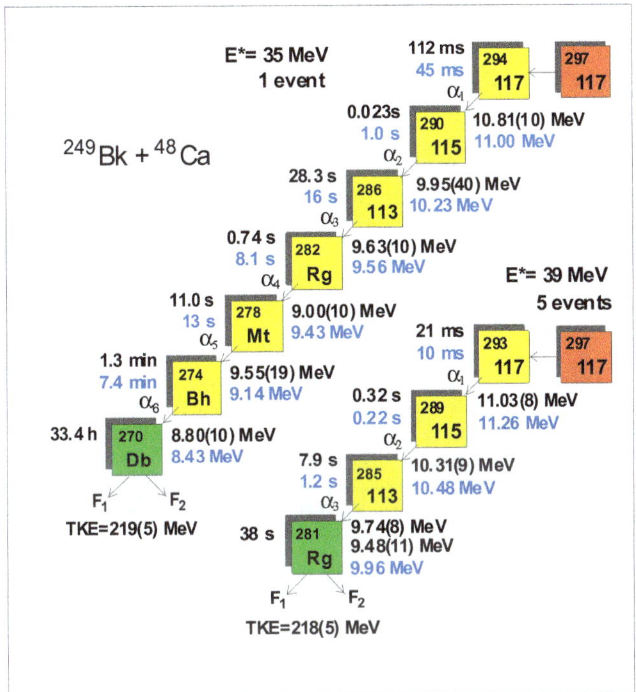

Fig. 5.

events of 294 and 11 new events of 293, to give 4 events of 294 and 16 of 293. GSI confirmed our discovery. With its half-life the order of 100 ms for 294, we had landed on the shore of the "Island of Stability". The observed half-lives of 112 and 113 are the order of 0.6 ms, 100 times shorter. From its time of production through the time of our experiment, ^{249}Bk is decaying to ^{249}Cf which bombarded with ^{48}Ca gives $Z = 118$, so we saw 1 more atom of 118. Hamilton reported for the collaboration the first report of the discovery of 117 at an international conference, the every three-year *Nuclear Physics Conference* covering all areas of nuclear physics in June 2010, in Vancouver, Canada. See Fig. 6.

Then, we repeated the experiment to produce elements 113 and 115. We obtained many more atoms of each. Before we could publish, the new International Union of Pure and Applied Chemistry and Physics committee asked us to submit all our published evidence for the new elements by

Synthesis of a New Element with Atomic Number Z=117

J. H. Hamilton[1], Yu. Ts. Oganessian[2], F. Sh. Abdullin[2], P. D. Bailey[3], D. Benker[3], M. E. Bennett[4], S. N. Dmitriev[2], J. Ezold[3], R. A. Henderson[5], M. G. Itkis[2], Yu. V. Lobanov[2], A. N. Mezentsev[2], K. J. Moody[5], S. L. Nelson[5], A. N. Polyakov[2], C. E. Porter[3], A. V. Ramayya[1], F. Riley[3], J. B. Roberto[3], M. A. Ryabinin[6], K. P. Rykaczewski[3], R. N. Sagaidak[2], D. A. Shaughnessy[5], I. V. Shirokovsky[2], M. A. Stoyer[5], V. G. Subbotin[2], R. Sudowe[4], A. M. Sukhov[2], Tu. S. Tsyganov[2], V. K. Utyonkov[2], A. A. Voinov[2], G. K. Vostokin[2], and P. A. Wilk[5]

[1]*Department of Physics and Astronomy, Vanderbilt University, USA*
[2]*Joint Institute for Nuclear Research, RU-141980, Dubna, Russia Federation*
[3]*Oak Ridge National Laboratory, Oak Ridge, TN 37831, USA*
[4]*University of Nevada Las Vegas, Las Vegas, NV 89154, USA*
[5]*Lawrence Livermore National Laboratory, Livermore, CA 94551, USA*
[6]*Research Institute of Atomic Reactors, RU-433510 Dimitrovgrad, Russia Federation*

INPC 2010, Vancouver, Canada

Fig. 6.

the end of May 2012. To be able to send our new data on 113 and 115 data, Hamilton presented their new data for the collaboration at the *Nucleus-Nucleus Collision Conference* in May 2012, in San Antonio, TX. See Fig. 7.

After nearly four years of deliberations, the committee informed us that we were given credit for the discoveries of elements 115, 117, and 118 and Japan for 113. We were asked to submit our names. In a Conference call including three from Dubna, Roberto ORNL, and Hamilton Vanderbilt, the Russians proposed Moscovium symbol MS for 115 after the Moscow region (State where Dubna is located), Hamilton's proposed Tennessine for 117 symbol Tn as he had gotten approved before the experiment if it was successful, and with Oganessian absence, the Russians proposed Oganession for 118 symbol Og. These were accepted by all and submitted. The Committee changed the symbol for Tennessine (the line ending is for 117 being in the column with fluorine, chlorine, and so forth) from Tn to Ts because around 1907 a new element was reported and named Thoron with the symbol Tn. It was later found to be an isotope of a known element but they wanted to avoid any confusion. See Fig. 8. The State of

New Insights into Discoveries of Elements 113,115 and 117

J.H. Hamilton[1], Yu.Ts. Oganessian[2] and V.K. Utyonkov[2]

[1]Vanderbilt University
[2]Joint Institute for Nuclear Research
And Collaboration with Oak Ridge National Laboratory,
 Lawrence Livermore
National Laboratory and Research
Institute of Atomic Reactors

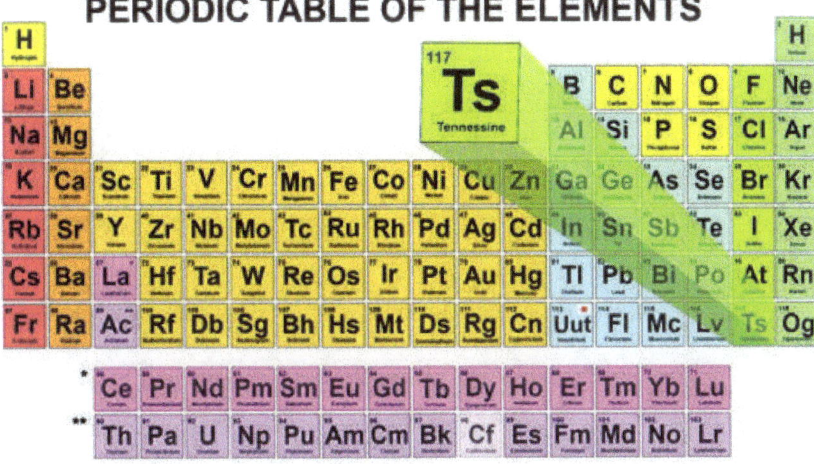

Nucleus-nucleus collision conf.
May 2012, San Antonio

Fig. 7.

PERIODIC TABLE OF THE ELEMENTS

Fig. 8.

Tennessee is the second state of the United States to have an element name for it and seen in periodic tables all around the world.

We want to acknowledge all the scientists involved in these discoveries as shown in Fig. 6 and as seen in subsequent publications submitted to the committee.

Teaching and Outreach of J. H. Hamilton

Sourish Dutta

Department of Physics, Vanderbilt University, Nashville, TN, USA

I have been his assistant in setting up demonstrations and carrying them out for the last several years. So, I have firsthand observed his teaching and he has shared many discussions with me.

Joseph Hamilton has had lifelong passions for teaching and outreach as well as research. His teaching has spanned all levels from beginning physics for engineers, to premeds and science students, to non-science majors, freshman honors seminars, upper-class seminars, courses for sophomores-juniors, seniors, and first-year graduate students to graduate students. For over forty years, he has taught at least one semester each year physics for non-science majors. First, look at a statistical list of his remarkable accomplishments followed by a discussion of each.

1. Career Highlights

1.1. *Teaching*

Taught over 10,000 students, giving over 6,000 lectures.
Developed many teaching demonstrations to teach physics.
Bed of nails, rocket cart, ultra-violet lines, and magnetic brake.
Summer science for 30 teachers and students for 12 years.
Gave 6 physics lectures to 65,000 Titans football fans plus those on TV.

2 15-min lectures on electricity and states of matter on local and national public television 8–10 times/year for 15 years.

Directed the honors research thesis of one high school senior and 12 undergraduate seniors.

1.2. *Research*

Over 1,200 publications in journals and books in 11 languages.

Directed/co-directed 64 + 10 (Chinese) PhD, 30 + 7 (Chinese) MS, 12 senior honors, 1 HS honor, and 10 summer research experience for undergraduate students.

Over 120 postdoctoral and research fellows from 20 countries.

Organized and chaired 13 international physics conferences.

Over 450 invited talks at international and national conferences.

He considers the physics course for non-science majors, one of the most important courses taught. This is because these students will go into all areas of life where they will be called upon to understand and judge the importance of scientific discoveries that impact their lives and to critically analyze the claims. So, they must know how science works and be critical in their judgments.

His course meets the core curriculum requirements and has become very popular mainly because of his extensive use of demonstration. He often asks students to join him in doing the demonstrations. Many students take pictures of the demonstrations. The Academic Councilor for student athletes reported how often she is shown pictures of his demonstrations. One professor said to him, "I watched you teaching a few times and I realized the difference in my teaching engineers. I occasionally use a demonstration to make a point, but you use demonstrations to teach physics." He does many more demonstrations than any other professor and has designed many of the ones he does himself.

Joe every year tells his class this story to emphasize how demonstrations teach physics so they are important for them to be attentive.

One year the men's basketball team invited him to go with them to their game at the University of Georgia as an honorary coach and sit on the bench. It was a very exciting game which Vanderbilt won by two points. A senior student who had taken his class earlier scored 22 points including the game winner. After the game, Joe was standing by the bench when the sister of this student came up to him and said I want you to meet my mother and father (she was now a student in his class). She said in introducing him to her mother, this is the professor who does all the neat demonstrations in class. Her mother said like what? She replied, "Last week he laid down on a bed of nails had his assistant place another bed of nails, nails down, on his chest and placed a 60-pound steel plate on the top bed of nails. Then he placed a board on top of the plate and with a hammer drove a nail into the board." Her mother said what did that show. To which she immediately answered, Newton's Law of inertia said the steel plate had so much inertia that it did not move to push the nails into his chest. Then Joe tells his class, if a student who had just watched her brother win an exciting basketball game could immediately describe a demonstration from the week before and tell the physics behind it then demonstrations do teach physics so observe them carefully.

He dresses up like Billy the Kid for his demonstration to teach how it is not just the force but the force times the time it acts that determine the momentum transfer. He used Civil War long and short barrel rifles and a Colt 41 pistol as used by Billy the Kid for illustrations. A Colt 45 had too much recoil for the Kid's small hands (no real bullets for them). His

Department Chair wrote the following in his retirement acknowledgment at commencement after describing his research accomplishments. "He also has a love for teaching--mixed with more than a little showmanship. Whether showing up for class dressed as Billy the Kid, lying on a bed of nails, or riding a fire-extinguisher-propelled cart across the front of the classroom, Hamilton's demonstrations brought physics to life. Physics was definitely not boring for the more than 10,000 non-science majors who took his Physics for Non-science Majors course."

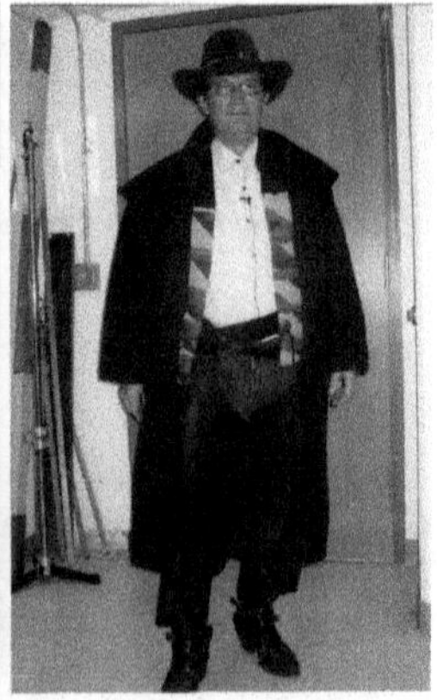

For his sophomore-junior level course, he wrote a textbook with a leading Chinese physicist entitled *Modern Atomic and Nuclear Physics*. The second edition was published by World Scientific Publishers who list it as one of their all-time best-selling textbooks.

A second reason for the popularity of his class for non-science majors is the following. Students are encouraged to take notes in class. They can bring their notes but not the textbook or electronic devices to the tests. His course does not involve memorizing lots of equations and definitions but emphasizes how to use these to solve problems. He stresses two important things they should learn — critical thinking and persistence. One of his PhD students nearly 60 years later wrote in her talk about how he was such an important influence in her life. "Joe Hamilton's philosophy 1. Hone your critical skills and 2. Be persistent".

Dr. Brantley in another section of this book describes his teaching and direction of graduate students and many of their accomplishments. With Joe, PhD research is very often a one-on-one teaching experience. There is one note I would like to add. Whitlock, one of his PhD students went to teach at Mississippi College, Joe's alma mater. One of Whitlock's students, W. Nettles, was sent to Vanderbilt to do his senior honors thesis with Joe. Then Nettles came to Vanderbilt and did his PhD thesis research under Joe. Nettles first taught in the Navy Nuclear Power school after being interviewed by Admiral Rickover. Later at Union University, he taught two students who went to Vanderbilt and did their PhD thesis research with Joe. Whitlock the son, Nettles the grandson, and Brewer and Eldridge the great grandsons. Very few if any can claim an amazing lineage of three generations of PhDs. The analogy of sons and grandsons is very appropriate because there is a very close family-like relationship between Joe and all his graduate students. Many have continued to do research with him and he has closely followed their careers. Just this last year he published papers in Physical Review and Nuclear Physics with a student who graduated over 40 years ago. At an AAAS awards ceremony for Joe, his DOE program research director said the most impressive thing he thought about Joe was how his students continually wanted to work with him for many years after graduation.

In 1978, he was the first American physicist to visit China's major nuclear laboratory far in the interior. He gave 3 days of lectures on frontiers in nuclear research there. He did the same again in 1981. Both sets of lectures were translated into Chinese and published there. They were used to train Chinese graduate students for over a decade in the frontiers of nuclear physics. In 1979, he published the first research publication in nuclear physics in a Chinese journal since before 1949. He brought the first two young Chinese research scientists in 1979 to the US and Vanderbilt. He had the first Chinese PhD student in nuclear physics who graduated in 1984. When he went back to China, he helped organize a joint PhD program where the Chinese students would take their course work in China, take the Vanderbilt written qualifying exams for the PhD, and if passed they would go to Vanderbilt for their PhD research with Hamilton who was appointed an honorary professor at Tsinghua University in Beijing. A similar program was set up with Fudan University in Shanghai. These are two of the leading universities in China. Ten Chinese students received the PhD and one Masters from Vanderbilt under these two programs. During this time, a Tsinghua professor came to Vanderbilt for a year and joined Hamilton's group studying neutron-rich nuclei produced in the fission of ^{252}Cf. When China began to develop their own PhD programs, Professor Zhu let them study new isotopes produced in this fission for a major part of their PhD thesis now at Tsinghua. Ten students received PhD and seven Master's degrees from Tsinghua through this joint collaboration with Prof. Hamilton.

In 1990, he developed a Summer Science Collaboration Program to help high students and teachers learn about exciting careers in different areas involving physics. The program shown in the following figure consisted of lectures and then doing a short experiment in the different laboratories at Vanderbilt from nuclear physics to using the Positron Emission tomography PET facility in the Medical School to identify where a radiation source had been placed in a loaf of bread. Then at Oak Ridge National Laboratory, they saw major research facilities and at Oak Ridge Associated Universities at their radioactive training laboratory used gamma ray detectors to identify the radioactive isotope that had been produced in the first atomic bomb explosion in New Mexico in fused sand. One teacher said he had been to two programs at National Laboratories and the Vanderbilt program was far superior. The program ran for 12 years and involved 240 students and teachers. One group is shown in the following figure.

Professor Hamilton has done many other outreach programs, For over 40 years he has given lectures to area elementary, middle, and high school students with demonstrations on What is Physics. National Geographic did a feature article for their Children's magazine showing him doing demonstrations. He was appointed the first Professor in Residence to give lectures to science students at Hume Fogg High School, Nashville's premier high school. Following his lecture with demonstrations of What is Physics to the whole student body, they stood up and gave him a standing ovation. The Principal said no speaker had ever received such an ovation.

Many years ago, he was asked by Nashville Public Television to do two 15-minute talks and demonstrations on Electricity and Magnetism and one on States of Matter. These were shown on Nashville Public TV during children's hours after school. These were sold to National Public TV and shown Nationally 8–10 times a year for 15 years. Professor's friends from Oregon to Boston called him telling him how much their children had enjoyed his presentations.

Early in his career he went to smaller colleges in several states to give a week of lectures on nuclear physics and help with the traveling radiation detection laboratory sent around by Oak Ridge Associated Universities. The program was to help faculty and students learn about nuclear radiation and its detection to help broaden their curriculum.

Joe was asked by the Titans to film six videos of the physics of football where they would add scenes from Titan players in actual games illustrating the physics principles, for example, quarterback Steve McNair throwing long spiral passes using the conservation of angular momentum which Professor Hamilton had just explained or Eddie George illustrating power running. These were shown on the Vanderbilt Jumbotron during time outs as shown in the following figure to 65,000 fans at the games plus many more viewing via TV. Joe used some of these in class every year to illustrate the principles. One of the current members of the Titans, Zack Cunningham, took Joe's non-science majors course. Athletes from many different sports took his course because of the demonstrations. He has taught many football players including over 40 who played in the National Football League. He taught all 14 members of the 2012 men's basketball team which beat Kentucky to win the SEC championship when UK went on to win the National championship. One member won 2 NBA championships. He taught 26 of the 28 members of Vanderbilt's NCAA National baseball championship team in 2014, to which he went. One of the teams who took his course became a star pitcher for the Los Angeles Dodgers,

Walker Beuhler. He taught Brendt Snedaker who won the 10 million dollar FED EX cup one year. Many women basketball, soccer, bowling, and tennis players took his course.

In recent years, he has given twice a six-lecture series with demonstrations in the Osher Lifelong Learning Institute at Vanderbilt with each lecture on a different area of physics for adults. His classes were among the largest they have had. Last year he was the featured speaker describing his work on the discovery of the new element 117 and naming it for the State of Tennessee, Tennessine, with symbol TS, at the Tennessee High School Science Teachers convention.

Next is a list of his Vanderbilt awards. One needs special mention. The Editor of the Vanderbilt yearbook posted announcements around campus for students to vote for their favorite outstanding professor. When the names were totaled, the Editor called Professor Hamilton and said his name had been written down twice as often as the next closest professor at Vanderbilt and they wanted to feature him in the Yearbook which they did as shown in the following figure. In another instance, his daughter who went to Alabama was in a group of Vanderbilt students and was introduced, one man said "You mean your father is Hamilton, Hamilton the Great".

Demonstrations Are Hamilton's Forte

Tired of boring lecture classes? Want something new and exciting? If so, then take Professor Joseph Hamilton's Physics 110a-b class. Although it is a large lecture class, Hamilton manages to enliven this PLE requirement course with his unique physics demonstrations. While explaining complex concepts, these demonstrations function also to break the monotony of class.

For Professor Hamilton, these demonstrations can do more than merely explain the properties of Physics. He uses them as a way to relate to the students. While they illustrate principles, they "break down barriers between professors and students." His demonstrations, such as his favorite cart and bed of nails tricks, receive many laughs from the undergraduates. Oftentimes, the students are surprised at the measures Hamilton will take to explain a principle.

Joseph Hamilton, a specialist in nuclear physics, is not only a teacher for graduates and undergraduates. He also actively researches and has been greatly recognized for it. This Sutherland Prize and George B. Pegram Award recipient is presently studying nuclear structures in hope of forming a unified nuclear model. His excellence in research and scientific leadership has won him prestige and influence. In 1971, Hamilton raised money in order to organize UNISOR, University Isotope Separator. This instrument was first used by university students to study the structure of nuclei. Hamilton also founded the Joint Institute for Heavy Ion Research. This one million dollar building is now used by numerous graduate and undergraduate students. However when asked if he would ever be interested in only doing research, Hamilton adamantly refuses because he thoroughly enjoys teaching, especially Vanderbilt's bright undergraduates. He comments that he likes teaching the nonscience majors course, Physics 110. Physics 110 has a long range impact by helping the students "understand the possibilities and limitations of science today and in today's society." Leslie Chalker, a sophomore in Peabody College, liked how Hamilton "always tried to show how Physics could be applied to our lives."

An example of this application stands out in Hamilton's mind. One semester, one of his students, Becky Scott was taking karate lessons. She had her instructor come to class to demonstrate the concept of impact by breaking cement blocks with his hand.

Outside of research and the classroom, Professor Hamilton relaxes by playing golf, shooting pictures (his wife complains that there is not enough space in the house for all his photographs), and traveling. Luckily, Hamilton has many opportunites to travel abroad through invitations to give lectures and attend seminars. His career in Physics has enabled him to visit sixty different countries. By being an adjunct and advisory professor at two universities in China, Hamilton and Vanderbilt graduate students are sent over to the Far East to develop programs and to do research.

Joseph Hamilton is devoted to his work. His incredible knack for Physics is easily seen through his lectures. Not only can he explain concepts such as force and motion, he can also demonstrate them with everyday materials. How much more exciting can a Lab Science course get?

ERIC STURBANK

In addition to lecturing, Joseph Hamilton also does many demonstrations. One of his favorites consist of his being sandwiched by a bed of nails. He even is brave enough to have a student stand on top. However, he can not forget the turban because, "it is very dangerous to lie on a bed of nails without a turban."

ACADEMICS 127

Vanderbilt Awards
Branscomb Distinguished professor
Sutherland Prize for Research
Furman Award for excellent teaching
Jefferson Award for service

Nordhaus Award for undergraduate teaching
Voted by students as Most Outstanding Vanderbilt Professor and featured in the annual in 1989 as shown on the right.

International Honors and awards for teaching, mentorship, and research
Humboldt Prize for Research, Germany
Ilkovic Gold Medal for Career Achievements, Slovak Academy of Science
Flerov Prize for Research, Russian Academy of Science
Award for International Scientific and Technological Cooperation with China from PRC President
International Cooperation Award, American Association for Advanced Science
American Physical Society Award for Mentoring
Southeastern section APS, Beams Awards research, Pegram Award Teaching, Slack Award Service, the only person to receive all three awards
Outstanding Professor of the Year for Tennessee, 1991, from Council for Advancement and Support of Education, Washington DC, one of four finalists for the national award
Resolution by Senate, House, Governor of State of TN honoring long career and discovery of new element 117 and naming it for the State of Tennessee
Similar resolution from Metropolitan Council, Mayor Nashville
Featured in the U.S. House of Representatives, congressional record for discovery of element 117
Eight honorary Doctors Degrees, St. Petersburg State U., Joint Institute Russia, Mississippi College, Eastern Kentucky U., Berea C., U. Frankfurt, U. Bucharest, Shukla U. India

National and international awards are listed above. He was recognized as the Outstanding Professor of the Year for the State of Tennessee for one year. He was among four finalists featured in the organization's journal for the Outstanding Professor of the Year for the United States.

In summary, Dean Geer of the College of Arts and Science in his Zoom remarks said "He is a legend at Vanderbilt." Indeed, no one has taught so many students and done so much research for 64 years on the faculty of Vanderbilt University.

Dr. Joseph H. Hamilton: Teacher, Research Advisor, Mentor, Colleague, and Friend

William H. Brantley

Department of Physics, Furman University
Greenville, South Carolina, USA

The first time I met Dr. Joe Hamilton was on a day in 1961 when I walked into his spectrometer lab in Learned Hall looking for him. The room was large and two-thirds of it was filled with a sprawling "machine" that one could go around, into, over, and under. In one corner next to a window, behind some shelves, there was a desk with students (who turned out to be several members of "The Big Team") looking over the shoulder of another student (I thought) sitting at the desk. The group paused in what they were doing, so I asked them collectively whether they knew where I could find Professor Joe Hamilton. I recall Herman Boyd backing up and pointing down at the "student" sitting at the desk, indicating "Here is Professor Hamilton," in a manner as though to say, "Don't you realize who this is!" Well, how was I to realize it? The person sitting there looked like the youngest student in the group. (He is still "one of the students in a group.") That meeting began for me a long and gratifying relationship with Dr. Hamilton.

There are many instances of Dr. Hamilton's teaching, some formalized in a classroom, others informal occasions of instruction about some aspect of the research we were doing. Physics is both difficult to learn and difficult to teach. It requires great effort of preparation and often extra effort of tutoring and nurturing. One of the first classes I had in graduate school was Electricity and Magnetism, taught by Dr. Hamilton. He was firm in letting us know what he expected, but he was also patient and helpful, always encouraging. He motivated his students. If we had problems digesting (as who would not?) the intricacies of J. D. Jackson's *Electromagnetic Theory*, he would help us work through problems that clarified the underlying concepts.

These experiences jump-started our learning and set us on track for eventual success in completing our graduate school requirements. Due to his caring and taking one-on-one time with me and others, Dr. Hamilton was a great model of how teachers should interact with their students and he was and is — still going strong — unique in the way he interacted with and *mentored* students.

My first research activity was learning to operate the "iron-free, double-focusing, beta-ray spectrometer of the Moussa Bellicard type," in all its various aspects. I worked with "The Big Team" (and eventually was added to it!) to help collect data for other students' theses. I knew how to do everything needed (including "oiling" its motor) and something of what the data might mean. For example, I had a clear understanding that it was not wise to leave the device alone for more than a few minutes at a time. Various disasters that could ensue, should something go wrong, had been carefully described to me. To take the long runs required for beta spectra data required several different people taking shifts.

One Friday about mid-day, Dr. Hamilton rushed through the lab door with his lab-coat-tails flying, a paper in hand, having obviously just come from reading the latest journals in the library. "Billy," he said, "finish what you are doing. I have a quick experiment I want you to run. These guys in this paper say there is no beta back-scatter with a source backing of a certain thickness, and I know they are wrong. They don't know what they are talking about. It won't take but a few hours to show that and since it is a short experiment you should be able to complete the data taking. You can use the Sodium-24 source. Let me see the data Monday morning." Dr. Hamilton helped me set up the experiment, and then he vanished as quickly as he had appeared.

I began taking data, thinking to finish up late that afternoon. Late afternoon arrived, and it was clear that the run would not be finished any time soon. As was well known then, each data point of a beta-ray spectrum taken on such a spectrometer could require much time, and endless numbers of data points were needed. I called my wife, Nancy, and said I was in the middle of something I couldn't leave and would not be home for a while. Late that night I told her it looked like I'd be there the rest of the night — I couldn't leave the machine and I wasn't near finishing. At one point, with some fear and trepidation, because the spectrometer was usually never left alone while running for more than a few minutes, I went to my apartment nearby to eat something and drink strong coffee, and then quickly back to the lab. Data collection continued through Saturday and into Sunday, by which time my lack of sleep was really catching up with me. The situation was saved to some extent by my dozing during the long-time counts on the down side of the positron spectrum. Finally, in desperation late on Sunday, I took a chance and set a long, long data point count. The run was almost done, and I was completely cooked! I had to take a chance and go home to sleep.

Early Monday morning I was back and the spectrometer was still intact and counting. Later in the day as I was completing the count on the last couple of points, Dr. Hamilton reappeared at the lab door. "Billy, what does the data look like? Let me see! Oh, yes, good! Have you ever done a Fermi-Kurie plot?" I'm thinking, "Fermi-Kurie who?" "No", I say. "Get some graph paper and let me show you how." In no time flat he's out the door, and I'm "Fermi-Kurie" plotting like a conspirator! As he left he told me to study certain books on Fermi's beta decay theory! As it turned out, that weekend and the following several weeks were an enormously instructive and productive time. We completed the analysis, wrote up a paper, and sent it off to be published soon after. I learned what F-K plots were all about and began seriously to learn beta decay theory to boot.

Although I had little sleep that weekend, I got the job done. I learned, as Dr. Hamilton intended, no doubt, that good work takes time and devotion, patience, and endurance. It was the best lesson he taught me, and it has carried me through my fifty-plus-year career.

But one other thing I also learned at the time was to be wary and ready in the future should I see Dr. Hamilton fast approaching and announcing a "short experiment!" I was all-in on whatever experiments presented themselves, but I did become aware of the necessity of enlisting the aid of

"The Big Team," for experiments that ran over days and days. Thank you to the Big Team — Barlow Newbolt, Walter Croft, Ted George, and Herman Boyd — and thanks also to you Ed Zganjar, Dana Willard, and Alice Hankla Davis, thank you all and thank any that I forgot to mention. Thanks to Dr. Hamilton for assembling "The Big Team" plus the other graduate students who were so helpful and welcoming to me as I began my studies.

"1963 Celebration of Dr. Joe Hamilton's Birthday with members of 'The Big Team' and others."

Joe and Jannelle Hamilton's home was always open to us. He invited us frequently to his house for social events where we got to know the other students. And his lovely wife, Janelle, always made us feel special and supported. It was a different time, one in which the university was a more collegial kind of think-tank, and Joe and his wife were its best exemplars.

Joe organized and chaired the *International Conference on Internal Conversion Processes*, at Vanderbilt University in 1965, which was a marvelous opportunity for all of his research students and post-docs. In an important way, this was like an extension of his research and teaching activities in his labs at Vanderbilt and his home. It was an unusual learning opportunity for his students, post-docs, and colleagues at Vanderbilt, and provided the opportunity to make new friends with colleagues from other parts of the world. Without realizing it, perhaps, Joe and Jannelle had well begun a lifelong activity of mentoring.

They extended the small dinners in their home for students, post-docs, and friends — with their conversations about physics and other topics of the day — to informal luncheons and dinners held at various regularly scheduled meetings of the American Physical Society and other research groups. With the growing numbers under his influence, and the faithful participation of so many, the small dinners became banquets. The international conference at Vanderbilt became the first of many conferences he helped to organize and direct. For example, the conferences 1st, 2nd, Etc., *International Conference on Fission and Properties of Neutron-Rich Nuclei (ICFN1-7)* became important in the world community of researchers in nuclear physics. Each of these conferences provided an opportunity for young and old across the years to meet and become acquainted or re-acquainted. Below are photos from some of these.

"Former students with Joe and Jannelle Hamilton at the Southeastern Section Meeting of APS in Auburn, AL, when Dr. Hamilton was given the Beams Award for Research in 1975."

"Joe Hamilton and a group of his students at the Fourth Sanibel Island, FL conference."

"Joe and Jannelle Hamilton with students, research colleagues, and family members."

1. Vignettes

One of the post-docs I worked with while a graduate student was Dr. A. V. Ramayya, who had just arrived at Vanderbilt from Indiana University. He liked to probe me and other students with physics questions from time to time to find out whether we actually knew anything. He mercifully spared us any "report card" but rather would use these interactions as teaching occasions. The last one of these interrogations occurred with Brooks Musangu, shortly after he arrived at Vanderbilt from Furman. He walked into Brooks' office one day and asked him what the half-life of a neutron was. Brooks responded that he could look it up. Dr. Ramayya said something to the effect, "I don't want you to have to look it up, I want you to know the answers to the questions I ask." He then went down the hall to his office and returned with *The Handbook of Nuclear Physics*, which he gave to Brooks saying, "get busy learning what's in here for the next time I have questions!"

At the end of the formal presentations in the June 8, 2022, International Zoom Meeting honoring Dr. Hamilton's life and work, a call was made to the Zoom audience inviting anyone who wished to make comments to do so. One of the responders particularly caught my attention — he was Dr. Suresh Pancholi, University of Delhi, a Post-Doc of Joe's with whom I had worked as a graduate student fifty-eight years earlier, in 1964. And many others also responded.

Not long after hiring me, Furman University hired another of Joe's PhD students, Ken Carter, for a tenure track position in physics. Ken did a great job during his relatively short teaching stint at Furman — short, I say, because Joe by now had been instrumental in starting a new research initiative at ORNL, namely, the UNISOR (later renamed UNIRIB) research consortium, of which Furman was a member. Ken and I went to an early UNISOR committee meeting in Oak Ridge, TN, the business of which was to hire staff positions to operate UNISOR. Ken was elected to one of the positions, and Furman lost an excellent teacher. However, UNISOR gained a great staff member, who in time became Director of UNISOR/UNIRIB. In this position he hosted and worked with researchers from all over the world who came to ORNL to do nuclear research, following — so to speak — in the steps of his major Professor, Joe. Ken had a long and distinguished career. He told me once that when he was the Director, he got an email from Joe in which he proudly announced, "This is my first email," which went on to discuss the business at hand. One wonders how many emails Joe has sent since then.

I began to do summer research with the UNISOR/UNIRIB consortium at the Holifield Laboratory in Oak Ridge. The lab is about ten miles from the town of Oak Ridge, TN, and if one were involved in a week-long experiment run, it was not easy to conduct the experiments and travel back and forth to motels in town. Joe Hamilton and others were instrumental in getting agreements and funding from federal, state, and local agencies for additional offices, conference facilities, and sleeping quarters near the Holifield Laboratory to remedy this situation. These facilities were used by local staff members and visiting scientists from all around the world.

During one of my summer stays at "the Joint," as it was affectionally referred to, I discovered that a member of "The Big Team" — Walter Croft — was also staying there. Thus, this facility enabled me to have the pleasure of getting re-acquainted with Walter and working alongside him again. It also enhanced the stay at the lab of many other visiting scientists.

Several years after I left Vanderbilt, J. B. Gupta, a master's degree-level teacher from the University of Delhi applied to do his PhD, degree under the supervision of Dr. Hamilton. There was a difficulty however — his university had only provided three years to complete the degree, and he had to leave his wife and family in India. Joe assured him that he could complete it, but pointed out that it would require work almost around the clock. With the benefit of long talks and much work with Joe, Dr. Gupta was one of only two who completed his degree in three years — quite an accomplishment — and he and Joe have very recently published their twentieth paper together.

Dr. Hamilton did not confine himself to working with college and graduate students. For twelve summers he and his associate director, Harold Crowell, ran a summer program to interest high school students in science. One-hundred-seventy students and 100 teachers went through this three-week program.

On a different note, in his thesis work with Joe, Crowell discovered that the lifetimes of excited nuclear states in heavy ion reactions could be measured. Dick Diamond from Berkeley built on Crowell's work and a floodgate opened.

In addition to Joe's work with graduate students and post-doctoral visitors, he also formed collaborations with other researchers.

A collaboration with two groups was formed in about 1990: one group represented by Shengjian Zhu, who was a visiting scholar to Vanderbilt in 1987–1989 from Tsinghua University, Beijing, People's Republic of

China, and the other group represented by Yuri Oganessian and Gurgen Ter-Akopian, both of the Joint Institute for Nuclear Research, Dubna, Russia. Ter-Akopian was a visiting scholar at Vanderbilt in 1990 and 1992. The focus of the collaboration was to study new neutron-rich nuclei produced in the spontaneous fission of ^{252}Cf and, together, the groups published several hundred research papers. Professor Zhu used these data as part of ten of his PhD students' theses and seven MS students' theses at Tsinghua University, and many of Joe Hamilton's students used these data for their theses.

Joe has described the relationships he often developed in working with others as akin to being a parent, and as the relationship of parent and child continues for a lifetime, so also his mentoring relationships have continued throughout his life. I experienced a prime example of this several years ago when it happened that I was asked by a teacher at Furman University's Osher Lifelong Learning Institute to give a lecture in his class on "the discovery of the new element, 117," which had not even been named. I agreed to do this. Of course, I knew that Joe and his group had discovered 117, so I called him to get help on some of the details of the discovery. Joe immediately sent me one of his lectures on the subject, complete with slides. Still, there was much work remaining for me to get up to speed on the subject matter, and in a subsequent discussion with Joe concerning some details, he simply volunteered to give the talk himself. I jumped on that offer "like a duck on a June bug," and that is how a rather large audience of retired learners at Furman heard about the discovery of element 117 from the discoverer himself!

Earlier I mentioned that Joe's last PhD student at Vanderbilt, Brooks Musangu (PhD 2021), was also the author's (PhD 1966) undergraduate research student in the physics department at Furman University. To use Joe's description of mentoring students as being akin to a relationship of parent and child, Brooks would be his "grandchild." Applying this idea to others of his students, Joe Hamilton taught Craig Whitlock at Vanderbilt. Craig Whitlock taught William G. Nettles at Mississippi College. William Nettles taught Nathan Brewer and Jonathan Eldridge at Union University. Each of these worked with Joe Hamilton at Vanderbilt, receiving their degrees as follows: Craig Whitlock (PhD 1969), William Nettles (PhD 1979), Nathan Brewer (PhD 2013), and Jonathan Eldridge (PhD 2020), or four generations of teacher-students, from Joe Hamilton to "child" to "grandchild" to two "great-grandchildren."

What an amazing gallery of students and post-docs there has been over sixty-four years. While I have mentioned only a few by name in my personal anecdotes, I think that everyone has received something by working with Joe. And everyone has contributed to the leader Joe Hamilton has become.

2. Contributions and Accomplishments of Vanderbilt Nuclear Graduate Students and Research Associates

Joe received his BS from Mississippi College in 1954 and MS and PhD from Indiana University in 1956 and 1958. Though Vanderbilt hired him immediately upon the completion of his work in 1958, the Dean graciously allowed him to honor a post-doc position that had been offered him in Sweden. He returned to Vanderbilt in 1959 and stayed until he retired in August 2022. In his 64 years of teaching at Vanderbilt University, he produced what must be a world-record 64 PhDs.

An important part of his great success is shown in his students, many of whom went on to long and distinguished careers of their own. His first PhD student, the late William Frey, taught for many years at Davidson College. I was his eighth PhD student, finishing in 1965 before going on to a post-doc in Delft and then to a still continuing career teaching at Furman University. And his last PhD student was Brooks Musangu from Lusaka, Zambia (who was, as destiny provided, my undergraduate student at Furman). Brooks finished his degree in 2021 with Joe and is currently on a five-year, post-doctoral appointment at Harvard University.

Of the 64 PhDs Joe mentored at Vanderbilt, nearly all went on to teaching and research in physics. Sixteen or so of Joe's students work for corporations, laboratories, and government agencies. And another 28 teach in universities across the US and the world. A quick snapshot shows his students located in some 19 states and six countries.

Three of Joe's students have been vice-chancellors and provosts; thirteen have been department chairs; seventeen have worked in Headquarters or Laboratories at the Department of Energy; two had careers at the National Science Foundation; another two in the Navy Nuclear Power Program; one was a professor at the Air Force Academy; and six spent part or all of their careers in nuclear medicine in hospitals and companies across the states.

A great number of Joe's students have been variously honored for their work.

Lee Riedinger, Professor and Vice President for Research at the University of Tennessee, Chair of the Division of Nuclear Physics, and Deputy Director of the Oak Ridge National Laboratory received the Slack Award for Outstanding Service from SESAPS, the South Eastern Section of the American Physical Society. So, too, did Edward F. Zganjar, Department Chair, Vice President for Research, and Alumni Professor of Physics at Louisiana State University. Likewise, Ken Carter, Director of UNISOR, recently won the Slack Service Award.

William H. Brantley, Amer Lahamer, and Craig Whitlock all merited Pegram Awards for Outstanding Teaching from SESAPS, and A. V. Ramayya earned the Beams Award for Outstanding Research from SESAPS.

Jerald D. Cole won two Edgerton, Germeshauser, and Greer, Idaho, Inc. (EGG Idaho, Inc.) $10,000 Awards for Outstanding Research in Nuclear Arms Control, and two IR 100 Awards for Outstanding Technical Developments.

Wayne Hibbits earned the Department of Energy Meritorious Service Award for contributions to environmental management at ORNL — the second highest award given to recognize outstanding employees, and Franklin E. Coffman, Former Deputy Assistant Director for Waste Management at DOE was Principal Investigator on two multibillion-dollar awards from the Federal Government: one for restoration in Iraq and the second to build a city for 30,000 people in Guam to house 10,000 Marines who had to leave Okinawa.

Finally, Christopher Goodin merited two awards for his research with the US Army Corps of Engineers.

Given the range of his students' important careers, Joe Hamilton's influence on American physics has been far-reaching and profound, to put it mildly.

But Joe was not only an influential teacher and leader. He was an excellent learner — and that was one of the secrets of his success. He learned not only by working with his graduate students but also by reaching out to and working with young post-docs and research associates. He invited to his lab any number of colleagues who had a different expertise or interest in the field than did Joe, and by 2011 he had hosted 165 separate visits — short and long — by 112 different scholars. His intellectual reach in this aspect of his career was also wide. The scholars he worked with represented 75 different institutes and universities from some 23 different countries. And their work together prospered.

Joe's collaborations helped him produce an average of one refereed publication every three weeks from 1958 until 2021, for a total of at least 1,114 scholarly research papers. (After his recent retirement, he was working on six more publications with colleagues.) And, of course, he brilliantly collaborated with scholars around the world on research that led to the discovery of not one but three new periodic elements — most notably perhaps Element 117, which he named in honor of his home state, Tennessine.

Those of us living in the mortal world wonder — no, marvel — at how Joe manages to keep all his activities going and to stay so positive in the face of occasional barriers he has faced in his work. I have one insight into that question, though it must remain only partial.

When I was still a young graduate student, Joe called me to his office to discuss various parts of my research. I thought we'd sit down and talk, but instead, I walked with him across campus to the Post Office and back. On the way over, I listened a lot. We walked at Olympic speed, and I had a hard time both breathing and walking. But not Joe.

And once he got his mail, and we were returning to the physics building, Joe continued to talk with me about my work — in depth and also at great speed. And as we talked and walked (still at sportsman's speed), he was opening, reading, and sorting his mail.

His multitasking was impressive, to say the least.

Joe is the son of a Baptist preacher, and he no doubt listened and learned from his upbringing. He took to heart the lessons his father taught, but especially the old biblical message about Christ: "In your hearts revere Christ as Lord. Always be prepared to give an answer to everyone who asks you to give the reason for the hope you have. But do this with gentleness and respect (1 Peter 3:15)."

That is Joe in a nutshell. Always ready to give reasons for hope, Joe does so with gentleness and respect — not just in matters of faith but in his working with the not inconsiderable number of God's children to whom he taught the miracles of God's universe. If physics is, as some have said, the art of discovering and making God's drawing board for the universe clear for all to see, Joe is without question one of the Creator's most able draftsmen.

The Last Graduate Student

Brooks M. Musangu

*Department of Neurobiology, Harvard Medical School,
Boston, MA 02115, USA
brooks_musangu@hms.harvard.edu*

This contribution highlights the time spent as an REU and graduate student under the supervision of Dr. Hamilton and Dr. Ramayya. I had the privilege of being the last graduate student of Dr. Hamilton. Here I hope to summarize in a few words what many may already have spoken of Dr. Hamilton as a mentor.

1. Introduction

Prof. Hamilton and I met in the fall of 2014 when he came to Furman University hosted by his former graduate student Dr. William H. Brantley to give a talk on the newly found heavy elements, including element 117 — Tennessine. I was starting my sophomore year of college at the time. I expressed my interest in seeking a summer research fellowship in Dr. Hamilton's group after being impressed by his work and storytelling abilities. Dr. Brantley was overjoyed and proposed that I spend the next year taking more advanced physics and mathematics courses to prepare me for the work.

In the fall of 2015, I applied for a Research Experience Undergraduate (REU) program at Vanderbilt University, to which I was admitted. However, due to my status as an international student, I did not qualify for funding from the program. Thankfully, I was able to raise a portion of the funds needed from Furman University and Dr. Hamilton supplemented the remaining balance from his research grant.

When I joined his lab in the summer of 2016 for the REU program, I was afraid of letting Dr. Brantley and Dr. Hamilton down after putting their trust in me. But Dr. Hamilton together with Dr. A. V. Ramayya gave me lectures on nuclear physics to get me started on the work. This was very reassuring because I knew that they were interested in my success more than anything else. This changed my relationship with Dr. Hamilton as I began seeing him as a mentor and all the stories about how great he was at mentoring began to be a reality to me.

2. The Mentor

My experiences with Dr. Hamilton as a mentor are most likely not that different from what everyone has mentioned or will mention in this volume. But of all the memories that I have of Dr. Hamilton as a mentor a few stand out to me both during my REU program and when I returned as a PhD candidate in his lab.

2.1. *His keen interest to invest in others*

When he accepted me as a REU student, Dr. Hamilton exhibited a strong interest in me and, particularly, in my professional goals. He made this clear to me right away by asking me specific questions at the beginning of the program. He was always willing to make an investment in me, but I wasn't the only one. When one walks into his office, the images of the people he has impacted on the walls say it all. As a mentor, he found genuine joy in helping others. And the mystery to me was how he made time for everyone. I have no idea how he managed to give each one of his students the same amount of attention. More importantly, he cares more about maintaining and growing his professional relationship. Recently, I got married and I didn't mention it to him, but he reached out to me, very excited about the news, and asked me to share pictures and tell him more about my wife.

It is such care that made it possible for me to have such a successful REU program where I worked on the study of the level structure of ^{104}Mo.

This work resulted in finding soft chiral bands in ^{104}Mo. I in turn submitted an abstract on this work to the Nuclear Physics Division Meeting of the American Physical Society in Vancouver in October 2016. This work was selected to be presented at a special session for REU students. See Figs. 1 and 2. More significantly, I submitted an abstract on the new chiral bands to the *Sixth International Conference on Fission and Properties of Neutron Rich Nuclei* held on November 6–12, 2016, on Sanibel Island, Florida.

2.2. *His expertise*

When I returned to Vanderbilt University in the summer of 2017 as a PhD candidate, I found myself once more afraid and nervous about what my next few years would look like. Mainly because I had no idea what topics or questions would be interesting to pursue. Yet again, I was privileged because I was working under Dr. Hamilton. Even then, it is generally advised that one should not get a principal investigator (mentor) that is five or ten years ahead because they might be removed from where one is

Fig. 1. From left to right, Dr. Ramayya, Dr. Musangu, Andrew Thibeaut, and Dr. Hamilton at the *Nuclear Physics Division Meeting on the American Physical Society* in Vancouver in October 2016.

Fig. 2.　Presentation of the work done during the *REU Program at a Symposium held at Vanderbilt University* in August, 2016. From left to right, J. H. Hamilton, B. Musangu, W. Brantley, and A. V. Ramayya.

that they can't relate to one's situation and be able to provide accurate advice. This was not the case with Dr. Hamilton. His level of knowledge and expertise in the field is vast and not only that, but he also somehow managed to stay on top of the constantly changing field. As such, he was quickly able to point me in the right direction which led to a successful graduate career.

Furthermore, of the many things that I can say about his level of expertise, I found his ability to look at a level scheme (whether in a paper or on a presentation) and be able to make a comment about what looked correct and what didn't within a small window of time incredible.

2.3.　*His willingness to share his expertise*

Just as it is important to have a mentor who has a great wealth of expertise, it is important that the mentor is willing to share their expertise. The many collaborations and professional relationships which Dr. Hamilton has maintained over the years speak to this fact. His office was always open to anyone who had questions and needed advice concerning their work. I specifically was always touched whenever he stayed late to review my work and share his expertise with me not because he wanted my work to be done his way but because he genuinely wanted me to benefit from the hard-won wisdom he learned over the course of his career. See Fig. 3.

2.4. *His respectful attitude*

For someone as accomplished as Dr. Hamilton, he is a very humble person. In all my interactions with him, he never criticized me or anyone around me. Not because I was a great student and was doing everything right, no! He just had his ways of giving only constructive feedback. This,

Fig. 3.   Reviewing my work with Dr. Hamilton for the then newly discovered second hot fission mode in Ce-Zr pair. From left to right: J.H. Hamilton, B. Musangu, E. Wang, and J. Eldridge.

Note: See B. Musangu *et al.* Anomalous Neutron Yields Confirmed for Ba-Mo and Newly Observed for Ce-Zr from Spontaneous Fission of ^{252}Cf, *Phys. Rev. C* **101**, 034610 (2020).

I believe, is one of the many reasons we are all taking this time to honor his life and career. He also did not shy away from giving some tough love whenever needed. This usually showed up whenever deadlines were approaching and I was not close to meeting them, he would tell me, "Let's pick up pace on that score, Brooks". More significantly, he would sit by my desk and would work and review things together until he felt like we were out of the woods. I consider myself extremely fortunate to have met a mentor who provides constructive, compassionate, and straightforward feedback and does not shy away from being honest for fear of hurting my feelings.

2.5. *His parting advice*

When it came time to leave Vanderbilt University, Dr. Hamilton, his wife Jannelle, Dr. E. Wang, and I shared a meal together and I asked him for one more piece of advice as I transitioned into my new career. He quoted to me one of his favorite quotes, "Nothing in the world can take the place of persistence. Talent will not; nothing is more common than unsuccessful men with talent. Genius will not; unrewarded genius is almost a proverb. Education will not; the world is full of educated failures. Persistence and determination alone are omnipotent" — Calvin Coolidge. He added with a smile on his face, "Jannelle says I am too persistent that sometimes it can be a vice". I then understood one of his secrets to his success and I have applied this principle in my life since then.

3. Conclusion

I hold Dr. Hamilton with the highest regard and with the utmost honor. Through his numerous works, his legacy will endure. To highlight a few of the fantastic works that emerged from our collaboration.[1,2] But more crucially, his influence on me extends into areas of my personal life in addition to my professional life. He taught me things that continue to reflect in the way I approach challenges, communicate with others, and prioritize things in life. Furthermore, he has inspired many of us to strive for excellence through collaboration, determination, and perseverance. Dr. Hamilton has been an excellent mentor and has prepared us to continue in his footsteps. The challenge is now on us to do our part.

References

1. B. M. Musangu *et al.* Chiral Vibrations and Collective Bands in ^{104,106}Mo, *Phys. Rev. C* **104**, 064318 (2021).
2. B. Musangu *et al.* Anomalous Neutron Yields Confirmed for Ba-Mo and Newly Observed for Ce-Zr from Spontaneous Fission of ^{252}Cf, *Phys. Rev. C* **101**, 034610 (2020).

Mentorship, Respect, and Collaboration

Christopher Zachary

cjzachary@gmail.com

From the start in the summer of 2015 till the end of my post-doctorate in the fall of 2020, I detail my time working for and alongside Joe Hamilton as we sorted out the structure of ^{157}Sm and ^{163}Gd.

1. Introductions

Meeting J. Hamilton and A. Ramayya was such an unassuming event. I had contacted them regarding the possibility of my transferring to their lab from working on a master's in health physics. So, it was that I came to their offices at the Stevenson Center at Vanderbilt University and sat down for a cordial chat about possible work in their lab. From the very start, J. Hamilton made me feel more like a potential colleague than a student. Both made me feel like I would be more than welcome working in their lab and so it was that I selected where I would pursue my PhD without even considering further groups at Vanderbilt.

For the next six years, I would work alongside J. Hamilton and A. Ramayya attending lectures, both impromptu and scheduled, in nuclear structure and experimental physics. We would discuss experiments past and future as we continued to evaluate data already in hand. They were always ready to explain anything I had a question for and I never lacked for questions.

2. Joining the Lab

My first exposure to nuclear structure was attending the defense of J. Hamilton and A. Ramayya's then student Enhong Wang. At the time I had no clear idea of what a level scheme was, and for the entire duration of the presentation I remained lost. However, that did not deter me from going forward with work in J. Hamilton and A.V. Ramayya's lab.

E. Wang and I would go on to work side by side on numerous projects for the duration of my time in J. Hamilton and A. Ramayya's lab. Our first task together was getting the lab ready for three new undergraduate researchers who would be joining the lab a mere month after me. So, it was that we were busy clearing up old work and setting up three additional workstations. All the while J. Hamilton and A. Ramayya were teaching me nuclear structure as E. Wang familiarized me with our software tools and how to analyze data to make physical observations.

In no time three more were working alongside us and I was tasked with coordinating their work. J. Hamilton then taught me and the three summer research interns nuclear structure side by side. I never expected I could manage the tasks set before me with so little training; however, J. Hamilton was confident in how he had structured the lab and was unsurprisingly correct. With help from J. Hamilton and his newest colleague, E. Wang, the four of us managed to make significant progress on each of the structures we had been assigned to study, each presenting our work at conferences the following year.

3. First Talks

After a whirlwind of a summer learning side by side with H. Fryman-Sinkhorn, J. Marcellino, and W. Lewis, it came time for our first presentation. Hunter Fryman-Sinkhorn and I made our way down to Mobile Alabama for our first nuclear structure presentations at the 2015 SESAPS meeting.

Joe Hamilton and the rest of his lab had worked with us on our presentations and waere confident that we would do a good job. I am glad to say that on this occasion we did not disappoint. H. Fryman-Sinkhorn presented gamma-vibrational bands in ^{103}Mo while I presented new structures in ^{157}Sm. Both talks were well received, and it was an excellent experience for both of us young researchers from J. Hamilton's lab.

A vanishing year later, J. Hamilton and all of us were putting on the *6th International Conference on Fission and Properties of Neutron-Rich Nuclei* on Sanibel Island. J. Hamilton and A. Ramayya, with the help of Carol Soren, had put together an excellent conference and our group was joined that year by Brooks Musangu, then an undergraduate researcher but later as a fellow research assistant and doctoral graduate, Jonathan Eldridge, a former undergraduate researcher now research assistant and future doctoral graduate, and Andrew Thibeault, an undergraduate researcher.

Once again J. Hamilton added to the lab's retinue smoothly. By the end of the summer, we were all working well together and had new work to present. This year we had the additional opportunity to aid in organizing and running a scientific conference together, an experience none will soon forget, especially because among all the work that had to be done J. Hamilton made sure there was plenty of opportunity to see the island together and form lasting connections in the nuclear structure community.

As the conference went on I had my first opportunity to present a colleague's work and see the true quality of mentor J. Hamilton. This kind of task bears with it the challenge of understanding work one may not have been as deeply involved in. In short, it was a disaster. Yet though I dreaded how J. Hamilton would respond when he learned of how the presentation went, I was surprised once again. Joe Hamilton took my performance in stride without showing any sign of losing confidence in me as a researcher. This is one of the examples of how I learned from J. Hamilton to treat students as successful potential colleagues and they will not disappoint.

4. Qualifying

In no time at all it seemed my qualifier was upon me. Fortunately, J. Hamilton had helped me in selecting the topic and in preparation for the qualifier. Several times he assured me I had nothing to worry about, however it would not be till my defense that the message sank in.

On the morning of my defense, I was beside myself with uncertainty. Many aspects of nuclear structure had yet to settle in for me; there would be many further conversations with J. Hamilton and his former students, now colleagues, before I began to fully understand the phenomena we studied. At the time, however, I had little sense of confidence and was again baffled by J. Hamilton's belief that I would do well. Thus, I made the mistake of using my qualifier as a conversation piece rather than a presentation. I had fumbled again, yet J. Hamilton stood by his selection of students. He and my committee gave me specific directions on what topics they wanted to see me address. Then, only a few weeks later and with J. Hamilton's guidance, I presented before my committee a second time and passed my qualifying exam.

Joe Hamilton once again demonstrated his excellence in mentoring, focusing not on the mistakes I had made but on the proper correction of them. In this and what feels like countless other interactions J. Hamilton made me feel like I was already one of his colleagues even when I was still his student.

5. Collaborations

Among the numerous opportunities to work with collaborators of J. Hamilton and A. Ramayya, three distinct opportunities stand out. These allowed me to gain experience in experimental conception, design, and execution on the local and international scale. All were made possible by the extensive network of colleagues J. Hamilton had grown over the years.

In the fall of 2016, I worked with S. Paulauskas and the University of Tennessee nuclear structure group under R. Gryzwacz at the National Superconducting Cyclotron Laboratory. There E. Wang, N. Brewer, and I helped with the setup and operation of the VANDLE array. We all worked along side the UT group to help carry out the two week data campaign studying beta-delayed neutron emission.

The following spring of 2017 J. Hamilton arranged for me to go work at the Joint Institute for Nuclear Research in Dubna, Russia. There G. Ter-Akopian brought me into his research group where we worked to finalize designs for an auxiliary detector to be used in tandem with the Gammasphere.

For the six weeks we worked together, we considered a variety of arrangements and materials that could be used, settling confidently on the final design. When our time together at the JINR was over, I returned to the States though we would continue to work together toward seeing our chamber installed in the Gammasphere.

Barely a year later in the spring of 2018, J. Hamilton gave me the opportunity to participate in the *Spontaneous and Induced Fission of Very Heavy and Super Heavy Nuclei Workshop* at the ETC* in Trento, Italy. Joe Hamilton arranged for me to present in his stead elaborating on future work we believed would be possible studying fission of very heavy and super heavy nuclei. This opportunity was well removed from the conferences I had attended in the past and gave me a chance to see firsthand research in the earliest stages. It was a priceless experience.

6. Last Talks

I clearly recall the most salient piece of advice J. Hamilton gave me for my defense, "do not put anything on your slides that you cannot explain." After my qualifier, I had not quite learned my lesson and sure enough I felt I still needed to include everything I had touched upon in my research. Thus, as my final presentation as a student of J. Hamilton came to a close, a member of my committee asked about some minor details. I had to admit that on that account I did not know, and J. Hamilton kind-heartedly laughed as I had once again not heeded that most simple advice of not including things I know I do not know.

Following my graduation in spring of 2019, J. Hamilton and I had the opportunity to continue working together and along with A. Ramayya, E. Wang, J. Eldridge, and B. Musangu. We would continue to pursue the study of fission fragments from ^{252}Cf spontaneous fission with a new experimental chamber. Unfortunately, this experiment was not to be as 2020 began our time slot was yet to be settled only for any discussion of the project to be put to rest by the events of spring 2020.

Again J. Hamilton demonstrated his commitment to his young colleagues, keeping us all busy to the end as he aided us in finding our future positions during 2020.

7. Conclusion

In my time as a student and then colleague of Joe Hamilton, he mentored me on how to be an excellent experimentalist and member of the scientific community at large. Under his mentorship, I had the opportunity to also learn from numerous other brilliant minds in the field, A. V. Ramayya, R. K. Gryzwacz, K. P. Rykaczewski, B. C. Rasco, S. V. Paulauskas, E. H. Wang, G. M. Ter-Akopian, B. Musangu, J. M. Eldridge, N. T. Brewer, and many more. Joe Hamilton demonstrated firsthand the value of respectful collaboration and enthusiastic involvement in the greater community.

Joe Hamilton enthusiastically shared his understanding of physics, most memorably in the numerous demonstrations of physical principles he would give for his introductory physics course at Vanderbilt University and local grade schools. This mentality of physics outreach he instilled in all his students, encouraging us to help in demonstrations.

Above all, Joe Hamilton showed how to respect oneself as a researcher by respecting the efforts of his colleagues. No matter how many times we had to take a structure back to the drawing board following conversations with J. Hamilton he never failed to show us respect. Thus, we learned to respect ourselves even in failure and to always keep going, no matter how difficult.

Working With Dr. J. H. Hamilton

J. D. Cole

Retired
jeraldcole@verizon.net

I discuss Dr. Hamilton's career and how he influenced my life, professionally and personally. His 2022 retirement celebration at Vanderbilt demonstrated his impact on students and colleagues through the years. His role in projects that helped benefit Oak Ridge National Laboratory is discussed, including major experiments in both the US and abroad. In conjunction with the experiments, he organized conferences that focused on neutron-rich nuclei and his role in producing the super heavy elements at Dubna, Russia.

In March of 2022, I was able to attend a conference in Nashville to commemorate Dr. Joe Hamilton's retirement from Vanderbilt University. I joined many of his former students and colleagues in reviewing his 64 years of teaching and mentoring thousands of students and supporting physics research and education around the world. His own personal accomplishments are commendable, but the conference reminded me that Joe always worked to support his students by introducing them to colleagues in the US and overseas who could open global opportunities for them. His students were like family, and he included them in many of the experiments he conducted in the US and abroad.

1. As a Student

Joe Hamilton influenced my life as a professor, colleague, and, most importantly, friend. He always supported and promoted his students, and he did this for me throughout my career.

As an example, he facilitated my transfer to Technische Hogeschool Delft in the Netherlands to enable me to complete my doctorate. He worked with Dr. Bob Van Nooijen, Wim Lourens, and other physics faculty, who invited me to come to Delft. I subsequently moved to the Netherlands and received my doctoral degree after two years. During this time, I was able to travel to Max Planck Institute for Nuclear Physics in Heidelberg to plan an experiment, the Cyclotron Laboratory at Louvain-la-Neuve for the same purpose, and to attend conferences in Germany. These were opportunities to meet European scientists, who have remained friends and colleagues for years.

2. Improving the Laboratories

Joe Hamilton worked very hard to increase the capabilities at the national laboratories to benefit not only his students but also others. The first was the University Isotope Separator at Oak Ridge (UNISOR), which was operated as a collaboration of various universities through Oak Ridge Associated Universities. This was a tremendous success for his students and students at the associated universities as they could do decay experiments on new isotopes for their degrees. This facility was supported by three permanent staff members, allowing for reliable and well-controlled operation of UNISOR. UNISOR later became a critical part of the Oak Ridge Radioactive Beam Facility. His efforts had far-reaching influences.

As part of his continuing effort to support and improve UNISOR, Joe worked to get a helium dilution refrigerator. I accompanied him on a fact-finding trip to Louvain-la-Neuve, Belgium, where a group had one on their cyclotron's mass separator. We spent a week there and participated in an experiment with it. It was a productive trip and UNISOR got a refrigerator that was in use for a time, but it required too much staff effort to use regularly for angular correlation experiments.

In addition to UNISOR, he worked to get an improved Recoil Mass Spectrometer at Holifield Laboratory which was a powerful tool for in-beam spectrometry. Dr. Tom Cormier had developed and built such a device with a split electric dipole for the Nuclear Structure Lab at Rochester University in New York. The spectrometer was an innovative device that allowed the separation of recoils from the target without the need for stopping the particles in an ion source, as was the method for UNISOR.

Dr. Cormier and I designed the spectrometer; it was much larger than the Rochester device, and Dr. Hamilton was able to secure increased funding through the Department of Energy and Oak Ridge National Laboratory (ORNL). The spectrometer was used for several years until the closing of the Holifield Facility.

The Joint Institute for Heavy Ion Research was a collaboration between the University of Tennessee, ORNL, and Vanderbilt University that Joe Hamilton had a lead role in building. It provided not only a large conference hall but also staff offices and guest rooms for visiting scientists and students who were working at Holifield Lab. This allowed interaction between the scientists and students in a relaxed and informal atmosphere.

3. Fission Fragment Studies

He also used the existing Compact Array and Gammasphere in new ways by placing spontaneously fissioning sources of ^{252}Cf and ^{242}Pu at the center of the arrays and studying the fission fragments. The Argonne National Laboratory Gammasphere and ORNL Compact Array experiments on ^{252}Cf and ^{242}Pu resulted in over 200 peer-reviewed journal publications for the collaboration that did this work. I participated in authoring over 100. Approximately 150 new level schemes for the fission fragments were determined that provided nuclear structure and behavior information different from that obtained through chemical separation and beta-decay. Distributions of neutron multiplicity were determined and published along with multiplicity determined from isotope pairs.

4. Superheavy Nuclei

Some of the most notable of Joe's work was with Yuri Oganessian at the Flerov Laboratory of Nuclear Reactions at the Joint Institute for Nuclear Research in Dubna, Russia. Through ORNL, Joe obtained old ^{252}Cf sources and had the ^{240}Cf extracted to make a target for use with Dubna's calcium beam and a target of ^{249}Bk. The result was the production of several super-heavy elements, of which element 117 was named Tennessine (Ts) in honor of Joe and the work at ORNL.

These are a few examples of Joe's promotion of global collaborations and his involvement in bringing his students into this international circle. He worked with scientists at Helmholtzzentrum für Schwerionenforschung (GSI) with heavy ion reactions to produce new isotopes for study.

5. Introductions at Dubna

His work at Dubna also led to my introduction and work with Gurgen Ter-Akopian in obtaining pure ^{239}Pu targets for use in neutron bean experiments. My Idaho National Laboratory colleague Dr. Mark Drigert and I traveled to Dubna in 2003 to meet with Dr. Ter-Akopian and his group to establish a collaboration that resulted in work at ANL's Intense Pulsed Neutron Source (IPNS). This was the n-alpha work that occupied Dr. Drigert and me for several years.

6. Conference on Neutron-Rich Nuclei

Joe Hamilton hosted several conferences through the years, the most famous of which is Fission and Properties of Neutron-Rich Nuclei. The early ones in the series were held in St. Andrews Scotland, while more recent ones have been held at Sanibel Island, Florida. The conferences at Sanibel took place in November and drew a large participation from European colleagues. The conferences stemmed from the studies with the ^{252}Cf and ^{242}Pu sources mentioned above. Both of these efforts produced neutron-rich isotopes, whereas standard in-beam experiments produced proton-rich nuclei.

Fig. 1. Joe Hamilton (center), Wenchao Ma (left), and myself (right) at the Sanibel conference in 2016. Good times indeed!

7. Summary

Dr. Hamilton has had an outstanding career spanning many years. I have been fortunate enough to participate in a small but valuable part — specifically the improvements he brought to ORNL with UNISOR, the Recoil Mass Spectrometer, and the Joint Institute until Holifield closed. In the same spirit of help, he supported his students throughout his career, supporting them and introducing them to new contacts around the world. His work on the super-heavy elements is honored with the naming of element 117 Ts after the State of Tennessee. These are only a few highlights of his remarkable career that involved me, and it is a great honor to have played a part in them.

My Mentor: Joseph H. Hamilton

Wenchao Ma

Department of Physics & Astronomy, Mississippi State University, USA

1. Introduction

Dr. Joseph Hamilton is a nuclear physicist with world reputation, and he has made great contributions to our understanding of atomic nuclei. As his PhD student at Vanderbilt University, I would like to talk about two areas of his research: the shape coexistence and properties of neutron-rich nuclei, which I was involved in and have a deeper understanding.

2. Shape Coexistence

Dr. Hamilton received his PhD degree in 1958 from Indiana University and became a faculty member at Vanderbilt University. That was a golden time of nuclear physics. Jensen and Mayer won the Nobel Prize in 1963 for their work on the shell model, and Bohr and Mottelson were awarded the 1975 Nobel Prize for their work on the collective model, "for the discovery of the connection between collective motion and particle motion in atomic nuclei and the development of the theory of the structure of the atomic nucleus based on this connection". It was found that most nuclei are either prolate- (like a football) or oblate- (like a pancake) deformed rather than perfect spheres. The Collective Model has been very successful in describing a variety of nuclear properties, including motions of the whole nucleus such as rotations and vibrations. A simple example is the rotational bands of nuclei with an even number of protons and neutrons.

124   *Science and Humanity: The Extraordinary Life of Joseph H. Hamilton*

Their energy levels can be described by $E = [h^2/2F]J(J + 1)$, where J is the level spin and F the moment of inertia, similar to the rotational energy in rigid body rotation. Dr. Hamilton's research established that the nuclei may have multiple possible co-existing states, and different shapes associated with them, called the "shape-coexistence". This can be demonstrated by his work around 1980 on even-even Hg isotopes, as shown in Fig. 1.

Take ^{184}Hg for example, it has two excitation bands, one is built at ground state, another at an excited 0^+ state. The excited band follows the $J(J + 1)$ *pattern* of a rotational band, with smaller level spacings, suggesting a larger moment of inertia. The ground-state band has larger level spacings, suggesting a smaller moment of inertia. Detailed research revealed that the excited band is associated with a prolate shape, rotating about an axis perpendicular to its symmetry axis, while the ground state band has a small oblate deformation, close to a spherical shape. All Hg isotopes depicted in Fig. 1 exhibit the same feature which shows that the nuclei of atoms do not have a fixed shape like a rigid body. Rather, they may have different shapes and properties depending on the quantum states they are associated with. Dr. Hamilton's group studied other nuclei, for example in mass 70 region, and experimentally established that the shape

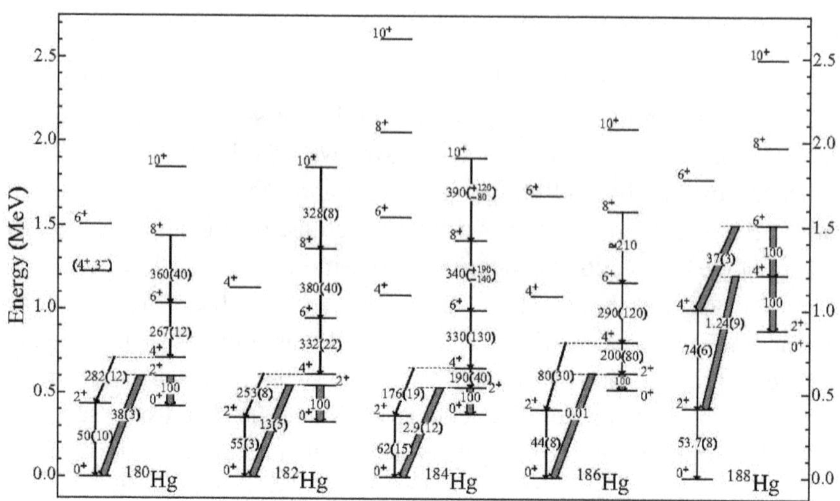

Fig. 1.   Excitation energy levels of light mercury isotopes.

co-existence is a general phenomenon in atomic nuclei. This is certainly a fundamental issue in nuclear physics. I personally participated in some experiments for this study and, in particular, worked on ^{182}Hg and ^{184}Hg as my PhD dissertation research project.

3. Properties of Neutron-Rich Nuclei

As shown in Fig. 2, the chart of nuclides, the line of stability deviates from the $Z = N$ straight line, bending towards the neutron-rich side. Thus, the combination of beam and target nuclei in the nuclear reactions always brings us to the neutron-deficient side. Before 1990, the nuclear structure studies were mostly limited to the neutron-deficient nuclei. A vast major-ity of neutron-rich nuclei could not be reached. There were a few heavy-ion beam fragmentation facilities in the world, where heavy-ion beam particles bombard a thin foil, and break into many fragments which tend to be on the neutron-rich side. The produced nuclei of interest need to be extracted from the fragments and analyzed. A small amount of

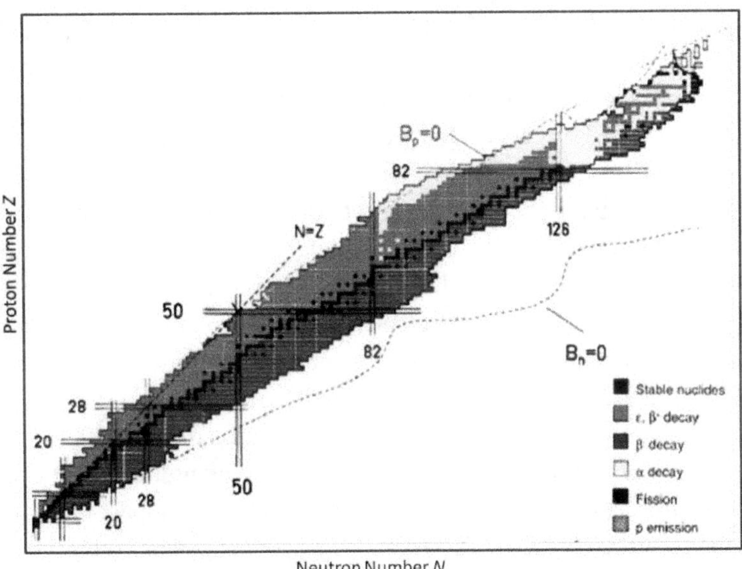

Fig. 2.   Chart of the nuclides.

neutron-rich nuclei can be studied this way. The production and measurements in these facilities are complex and expensive.

The spontaneous fission of ^{252}Cf can produce a large number of daughter nuclides which are neutron-rich. However, the fission data are too complex. In conventional spectroscopic studies, the reactions are always carefully selected so that the nuclei of interest are the primary products or, at least, from a relatively strong reaction channel, so that more decay gamma rays come from the nuclei to be investigated. The fission process produces many dozens of different fission fragments. The decay gamma rays of all these daughter nuclei are contained in the same data set. The data is thus very "dirty". When gating on a particular gamma-ray energy in the coincidence analysis, you see gamma rays from many different fission products. Furthermore, at least two fission fragments are produced together and their decay gamma rays are in coincidence. It is not possible to only see gamma rays from a single nucleus. Without mass/charge tagging, it is also very difficult to identify which nuclide the gamma rays belong to. Therefore, it was not possible to study the level structures of neutron-rich nuclei from the decay gamma rays of the fission products before.

In the late 1980s, large Compton-suppressed germanium spectrometer arrays were developed in the United States, such as the Compact ball at Oak Ridge National Laboratory (ORNL) with 36 spectrometers, and the Gammasphere at Lawrence Berkeley National Laboratory (LBNL) with 108 spectrometers. The high resolving power for gamma rays provided by these arrays brought the nuclear spectroscopy, especially the high-spin studies to a brand new stage. While most people concentrated on working on neutron-deficient nuclei, Dr. Hamilton considered using the high resolving power to study the neutron-rich nuclei from the fission products. He discussed with me in 1987 while I was a postdoctoral research associate at Argonne National Laboratory (ANL). In 1990, we carried out a measurement at ORNL with the ^{252}Cf fission source mounted at the center of the Compact ball. It was during the Thanksgiving holidays that nobody was doing experiments using the accelerator and the Compact ball. So, we got a longer time for our measurements. Later on, several measurements were carried out using the Gammasphere. A great amount of fission data was collected, which is the best fission data set in the world. The data analysis is highly sophisticated. One has to be extremely careful. Dr. Hamilton led the group performed three and four fold coincidence gamma ray analysis, and angular correlation measurements as well. The research project was highly successful. More than 150 papers were published in journal articles, and over 70 invited talks were given at international conferences. This work probably produced more results on the structures of neutron-rich nuclei than any research group using other approaches in the period of two decades of time.

Dr. Hamilton initiated and chaired the *International Conference on Fission and Neutron-Rich Nuclei* in 1997, which was held six times up to 2016 and a seventh being planned. Each time, more than one hundred nuclear physicists came from all over the world to discuss the frontier topics in the field. Dr. Hamilton did not stop here. He further studied the super-heavy elements. He was one of the discoverers of element 117, Tennessine, which was named after the state of Tennessee because of his contribution.

4. A Great Mentor

I would like to also mention that Dr. Hamilton was truly exceptional in mentoring hundreds of early career nuclear scientists, including undergraduate and graduate students, postdoctoral scholars, and nuclear science professionals. He directed more than 60 PhD and 30 Masters theses. He has also been the mentor for more postdoctoral research associates than those two numbers combined. Many of his graduate students have gone on to prestigious positions. More than a dozen or so have become chairs of various departments in physics or other areas, such as medical physics, and with as many going into research at national laboratories. He works intently with each one of his graduate students during their academic studies and while doing their thesis research, but he also maintains contact with them after their graduation.

I was his graduate student from 1981 to 1985 and a research associate between 1990 and 1993. So, I have personal experience with his mentoring activities. He helped me to learn and to grow and has given me guidance and help in every important moment of my life. During each experiment, he worked with us days and nights in the lab; back in the university, he sat with us side by side analyzing data whenever he had time. He is an exceptionally vigorous person and works highly efficiently. He thinks fast, talks fast, writes fast, and even walks like trotting. When I finished my 200-page PhD dissertation, he carefully read through it and made comments and corrections on every page, including English language corrections. We performed experiments at ORNL, ANL, and LBNL. We used the Gammasphere — Fragment Mass Analyzer (FMA) combination to push the limits of nuclear stability to the very neutron-deficient side in the mass 70 and 180 regions. He always encouraged me to look for new ideas, write research proposals, take more and more responsibilities in experiments and data analysis, and learn to supervise graduate students. In 1994, I took an Assistant Professor position at Mississippi State University and started to set up my own research program on high-spin physics. I had to also make a transition from using VAX computers to UNIX workstations. The help Dr. Hamilton gave me during this transitional period was beyond what I could imagine. We had several joint experiments in the first several years, where Dr. Hamilton helped me work on proposals and supported me to take the lead for the research projects. He even agreed to loan me one of his VAX computers so that my research would not be interrupted. All these efforts helped me quickly obtain a DOE research grant in 1995 which continued without interruption until my retirement. The aggregate amount of my grants and journal publications ranked highest among all faculty members in our Physics Department. I also obtained a Faculty Recognition Award on Research of the University.

Dr. Hamilton helped not only me but also our Physics Department at MSU. He served as an external reviewer in one of our five-year reviews and made a number of suggestions for the future development of the department. I am glad that we followed those suggestions and the department changed significantly, from a basically teaching service department to a department with substantial research activities, and from less than ten graduate students to more than fifty. Dr. Hamilton's help to me is unforgettable. I know that his other students all have the same feeling about his care, help, guidance, nurture, and love.

An important aspect of Dr. Hamilton's mentoring activity is through research collaborations. He is a highly respected nuclear physicist in the world and his scientific contributions have produced significant impact in the field. He has been responsible for a large amount of excellent scientific work of others through his organization of collaborations, consortia, facilities, and instrumentation to support these works. For example, Dr. Hamilton led the creation of the Joint Institute for Heavy Ion Research (JIHIR), the University Isotope Separator (UNISOR), and the University Radioactive Ion Beam Consortium (UNIRIB), as well as the construction of the Recoil Mass Separator (RMS) at the Holifield Radioactive Ion Beam Facility at ORNL. Hundreds of young graduate students, postdoctoral research associates, and junior faculty members benefited from working there closely with each other and with senior scientists. His graduate students and postdoctoral research associates came from more than 20 different countries. He has many contacts and collaborations with scientists all over the world, including those in European countries, Russia, and China. Many research projects could not be completed without collaborations of researchers in different countries.

https://doi.org/10.1142/9789811296963_0015

Reflections on Over 50 Years Working with Joseph Hamilton

J. B. Gupta

Ramjas College, University of Delhi, Delhi, India

I was a PhD scholar under Professor J. H. Hamilton in 1970–1973 and have continued to work with him for the last 50 years.

In the late 1960s, I was teaching physics at Ramjas College with a master's degree in physics. I needed to obtain my PhD to advance my career. To do this, I asked my University for a three-year leave of absence to obtain my PhD. They granted my request. I applied to Vanderbilt University to do my PhD thesis work with Prof. Hamilton and I was given a fellowship to do this.

I quickly met with Prof. Hamilton to ask if I could complete my degree in three years. He jokingly said yes by working 24 hours a day. I quickly completed my coursework and began my thesis research work with him.

It was a hard three years because I had left my wife and five children in India. Prof. Hamilton often met with me to encourage and cheer me up during these years as well as work with me on my thesis research.

I completed my PhD in the three years, one of only two of his PhD students to do so. In the summer of 1973, as I was writing my thesis, my wife came to be with me. When she first met Prof. Hamilton, she said "You cannot imagine how much love I and my children have for you for all you have done for my husband." Many years later when he came to India, we enjoyed having him in our home for dinner.

Following the publication of my PhD thesis work, I have continued to do research with him for 50 years co-authoring 22 research publications, with the last two publications in 2022. In 1984, he and Prof. Kumar supported me to attend an *International Workshop in Oak Ridge National Laboratory*. Beginning in 1997, Prof. Hamilton organized and chaired a series of six conferences on *Fission and the Properties of Neutron Rich Nuclei* with the proceedings published by World Scientific. Prof Hamilton provided major financial support for me to attend the first, third, fourth, and fifth of these conferences on Sanibel Island, FL in 1997, 2002, 2007, and 2012. I presented 2, 2, 3, and 2 papers (7 authored by only me) at these conferences and these were published in the proceedings as well as later in journals. I also gave a talk at APS conference in Nashville in 1997.

Prof. Hamilton continuously encouraged me in my research including my many papers without him.

Notes on the 20 research publications are listed as follows:

With the background of the extensive experimental research work of Professor Hamilton and his research scholars over the last sixty years at Vanderbilt University, the research work of J. B. Gupta in collaboration with Professor Hamilton represents the experimental and theoretical analysis of the large number of nuclei. It encompasses the empirical analysis, the calculations in the algebraic Interacting Boson Model of Arima and Iachello, as well as the calculations in the microscopic theory, based on the dynamic Pairing plus quadrupole (DPPQ) interaction in the Hartree Foch Bogoliubov (HFB) approximation, of Kumar and Baranger. The work covers the light nucleus of $Z = 30$ Zn, up to the well-deformed heavy nucleus of ^{168}Er:

1. The coincidence spectrum of ^{154}Gd was studied at Vanderbilt Lab. including the angular correlation. A theoretical analysis of the same was done in the Pairing plus quadrupole model in collaboration with Professor Krishna Kumar. The same was again analyzed in the DPPQ model and the interacting boson model. It serves as the first instance of a nucleus with 8 multiple collective bands (Refs. 1, 2, 3, and 12).

2. For the first time, the nuclei in the $A = 140$–180 region were grouped in 3 quadrants of the major shells. The formation of the neutron-dependent

multiplets in quadrant-1 and proton-dependent multiplets in quadrant-3 was illustrated therein. The F-spin multiplets were formed in quadrant-2. This division of the nuclear chart was linked to the beta stability valley. The nuclei in the three quadrants are associated with different spectral features. This division has helped in the proper and convenient view of the nuclei (Ref. 5).

3. From our extensive analysis of the $K\pi = 0_2^+$ bands, these were shown to be axially symmetric beta vibrational bands, as a rule, rather than as an exception, the view being held at that time.

4. A critical analysis of the nuclei, associated with the SU(3) symmetry, was presented in Ref. 19 of the following list. The problem of the separation of the beta and gamma bands was explained in the context of the SU(3) symmetry.

5. A detailed study of the collective multiband spectra of several chains of isotopes was presented in these publications, extending up to the six multiphonon bands.

6. A proper perspective of the expression of the kinetic moment of inertia, limited to the well-deformed nuclei was presented (Ref. 10).

7. Spectral features of the anharmonic vibrator nucleus $N = 88$ ^{150}Sm were successfully analyzed in the dynamic PPQ model of Kumar-Baranger, and later it was analyzed in IBM as well (Ref. 9).

J. B. Gupta is inspired and supported by Professor Hamilton in all these listed works.

A list of 20 research publications in the Refereed journals is given as follows:

1. J. B. Gupta, S. L. Gupta, J. H. Hamilton, and A.V. Ramayya, Gamma Gamma Directional Correlations and Coincidence Studies in Gd-154, *Zeit Phys.* **282**, 179–189 (1977).

2. J. B. Gupta, K. Kumar, and J. H. Hamilton, Pairing Plus Quadrupole Model Calculation for ^{154,156}Gd, *Phys. Rev.* **C16**, 427–437 (1977).

3. K. Kumar, J. B. Gupta, and J. H. Hamilton, DDT Based On The PPQ Model and Its Extension to Multi-Phonon Vibrational Bands (for ^{154}Gd), *Austr. J. Phys.* **32**, 307–322 (1979).

4. K. Kumar, J. B. Gupta, and J. H. Hamilton (VU Work), Proton Spectroscopic Factors of Sm Isotopes in PPQ Model, *Nucl. Phys. (Amsterdam)*, **A448**, 36–44 (1986). J. B. Gupta was hosted by Professors Hamilton and Krishna Kumar in (1984) for the International Workshop at ORNL.

5. J. B. Gupta, J. H. Hamilton, and A. V. Ramayya, A Comparison of F-Spin and NpNn Schemes with Global Systematic, *Int. J. Mod. Phys.* **A5**, 1155 (1990).

6. J. B. Gupta, H. M. Mittal, J. H. Hamilton, and A. V. Ramayya, Systematic Dependence of γ-g B(E2) Ratios on NpNn Product, *Phys. Rev.* **C42**, 1373 (1990).

7. J. B. Gupta, J. H. Hamilton, and A. V. Ramayya, Nuclear Structure of ^{158,160}Dy in the Dynamic Pairing Plus Quadrupole Model and the Shape Phase Transitions in $^{154\text{-}160}$Dy, *Nucl. Phys.* **A542**, 329–353 (1992).

8. J. B. Gupta, J. H. Hamilton, and A. V. Ramayya, A Microscopic Study of ^{168}Er Multiple Band Structure, *Phys. Rev.* **C63**, 04-4306 (2001).

9. J. B. Gupta, K. Kumar, and J. H. Hamilton, Nuclear Structure of ^{150}Sm, in the DPPQ Model and IBM, *Int. J. Mod. Phys.* **E19**(07), 1491–1511 (2010).

10. J. B. Gupta and J. H. Hamilton, Anomaly of the Moment of Inertia of Shape Transitional Nuclei, *Phys. Rev.* **C83**, 06–4312 (2011).

11. J. B. Gupta and J. H. Hamilton, Nature of $K^{\pi}{=}0_2^+$ Bands in A=140-180 Region, a Global Analysis, *Eur. Phys. J.* **A51**, 151 (2015).

12. J. B. Gupta and J. H. Hamilton, Re-examination of Nuclear Structure of ^{154}Gd in DPPQ Model, *Phys. Rev.* **C95**, 054303 (2017).

13. J. B. Gupta and J. H. Hamilton, Outstanding Problems in Nuclear Structure of ^{152}Sm, *Phys. Rev.* **C96**(03), 4321 (2017).

14. J. B. Gupta and J. H. Hamilton, Empirical Study of the Shape Phase Transition and Shape Coexistence in Zn, Ge, Se, *Nucl. Phys.* **A983**, 20–37 (2019).

15. J. B. Gupta and J. H. Hamilton, New Perspective of the Nuclear Structure of $^{96\text{-}114}$Ru, *Eur. Phys. J.* **A56**, 14 (2020).

16. J. B. Gupta and J. H. Hamilton, Shape Phase Changes with N in $^{72\text{-}84}$Kr, *Int. J. Mod. Phys.* **E29**(09), 2030008 (2020).

17. J. B. Gupta and J. H. Hamilton, Spectral Features of Nuclear Structure of $^{114\text{-}122}$Xe, *Eur. Phys. J.* **A57**, 97 (2021).
18. J. B. Gupta and J. H. Hamilton, Nuclear Structure of ^{126}Xe, *Phys. Rev.* **C104**, 05–4325 (2021).
19. J. B. Gupta and J. H. Hamilton, Analysis of the SU(3) Symmetry Versus Rotor Model, *Phys. Rev.* **C105**, 06–4312 (2022).
20. J. B. Gupta and J. H. Hamilton, Novel Features of Nuclear Structure of $Z = 46$ Pd Isotopes, *Nucl. Phys.* **A1028**, 122527 (2022).

Publications while he was a graduate student:

1. J. H. Hamilton, S. M. Brahmavar, J. B. Gupta, R. W. Lide, and P. H. Stelson, Search for Weak Transitions in the Decay of 108mAg, *Nucl. Phys.* **A172**, 139–144 (1971).
2. J. B. Gupta, N.C. Singal, and J. H. Hamilton, Search for Weak Transitions in the Decay of 125Sb, *Z. Physik* **261**, 137–142 (1973).

Joseph H. Hamilton's Influences on My Life

Reginald M. Ronningen

Senior Physicist FRIB/NSCL Emeritus
ronninge@msu.edu

I am honored to present this paper that documents Joseph H. Hamilton's attributes as an exemplary mentor, valued colleague, and lifelong friend. It begins with how, when, and where we first met. It then documents my recollections of events in my professional and personal life that show Joe's influences via his attributes as a mentor, advisor, colleague, and dear friend.

1. Introduction to Vanderbilt University and Joe's Scientific Group

I first learned about Vanderbilt University (VU) and its excellence from a graduate of the (*nka*) University of Wisconsin-River Falls (UWRF) who returned to teach there after obtaining an MS degree under Prof. Robert Lagemann. Nearing graduation from UWRF in 1969, I was considering a career in health physics and applied to the physics department at three universities. VU was the first to respond and offered me a teaching assistantship in physics, as by then, the Atomic Energy Commission fellowship program had been terminated. I accepted the offer, and in September 1969, I traveled to Nashville by Greyhound bus with $100 and two suitcases (only one arrived).

I did not at first "officially" join a research group. Instead, I took coursework and preliminary exams and performed lab work for a research associate. I almost wrote off joining Joe's group that first term. Incoming students were then required to visit each research group. The appointed times were on Thursdays. After a long week of classwork, my fatigue and a rather unenthusiastic presentation by several junior graduate students in Joe's group left me uninterested. But, after several years, Ken Baker and Eugene Collins from Joe's group asked me to modify a computer program that calculated M1 and E2 gamma-ray multipolarity mixing ratios to include an M3 contribution. This did interest me. Afterward, they asked me to consider joining Joe's research group. I sensed their deep dedication to Joe, and I did.

During my early days with Joe's group, he was often on the telephone when I would go to his office. Fortunately, Ramayya was usually always available to help me. I did not realize then that Joe was probably conversing with a senator, funding officer, UNISOR consortium, or another scientific colleague! Later, Joe and his office were busy but always welcoming.

2. Joe as a Mentor, Colleague, and Friend

Joe not only served ably as a mentor and esteemed colleague but additionally cared deeply for his students, especially in helping them continually improve their lives and scientific skills and promoting professional growth.

2.1. *On life skills*

For new group members, Joe and his wife, Jannelle, invited us to their home to bond socially. Early on, Joe sensed that I needed to boost my confidence level and suggested that I take a Dale Carnegie course. Probably due to a busy course schedule, I never took it. However, a Department of Physics and Astronomy support staff member did, and I witnessed that person's sincere expression of happiness and satisfaction at the completion ceremony. In hindsight, the Carnegie course booklet

Meetings: Quicker & Better Results would have been handy during my professional time later!

2.2. *On my scientific development as a graduate student*

One of my first tasks when in Joe's group was plotting (by hand) Coulomb excitation data that Larry Varnell and Jurgen Lange had taken using the scattering of ^{4}He from stable hafnium isotopes, i.e., using (α, α') reaction. The analysis was complicated by the target material being contaminated by tungsten; a tungsten wire was probably used to grow hafnium crystals, and the spectra contained peaks from the tungsten isotopes. This work was frustrating but proved helpful for the part of my thesis to analyze data taken using new, high-purity ^{176,178,180}Hf and ^{154,156,158,160}Gd targets,[1-3] made for Joe by Kelly Dagenhart at ORNL.

One of Joe's attributes is that he is a visionary. He had the foresight to also order other rare-earth-region nuclei ^{144,146,148,150}Nd, ^{156,158}Dy, ^{162,164}Er, ^{168,176}Yb, and ^{192,194}Pt for our future studies.[4-11] These studies had a significant scientific impact on my later career.

Another major part of my thesis was on measuring the lifetimes of the first excited 0^+ states in ^{72,74}Se. These data helped test a nuclear model explaining the coexistence of differing shapes in the same nucleus.[12-15] For example, the ground state of ^{72}Se is considered the head of a band of excited states characteristic of a vibrating sphere. However, the first excited 0+ state heads a band of states characteristic of a rotating deformed object. The analysis of the ^{72}Se data produced a spectrum of gamma rays from decaying excited states in ^{72}Se using a condition, or "gate," on a gamma ray that feeds the excited 0^+ state. Computer memory at the time limited the spectrum size to 1,024 data channels instead of a more desirable 4,096 or 8,192 channels. A critical gamma-ray doublet in this spectrum had to be resolved by hand, as, at the time, computer analysis would have been less reliable for resolving this doublet. Ramayya taught me "by hand" spectral unfolding approaches, including using appropriate peak shapes and a planimeter to obtain peak areas for relative intensities. Bob van Nooijen, who was visiting from the Technische Universiteit, Delft, and an expert at such analysis, verified my work. This "by hand" analysis was also used in analyzing my Coulomb excitation data; computer programs to analyze such data, where peak shapes significantly departed from "Gaussian," were also not generally available.

All of my lifetime and Coulomb excitation data were taken at the tandem Van de Graaf accelerator facility at ORNL. This work, with the support of Joe and talented collaborators, helped launch my career in nuclear science. To efficiently complete data collection, we had to learn how to keep the accelerator running overnight and on weekends. Also, we had to learn how to "tune" the accelerator beam. These skills served me well at other accelerator facilities, including those at Michigan State University (MSU). I'm honored that I am one of few who can say that they "tuned" the K50 cyclotron, both the K500 and K1200 superconducting-magnet cyclotrons and the superconducting-magnet cyclotron constructed for use in cancer therapy.

2.3. *A shape coexistence model recollection*

Prof. Tom Pinkston, who worked on the shape-coexistence model theory, was preparing to leave to visit Prof. Walter Greiner in Frankfort, Germany. On a day near his departure date, Joe received a letter from Prof. L. K. Peker containing his theoretical insights. It was written in Russian. Joe asked me to take it to VU's Russian language department to have it translated. I did so by the early afternoon, but no staff seemed present. After a long wait, an older professor arrived, and I asked if he would translate the letter. He stood silently reading it and then handed it back to me. He told me he would not translate it because its subject dealt with science and technology! On my way out of the department, I found a young faculty member expert in Slavic languages who could and did translate it. He was congenial and curious and even asked me to describe this physics. Joe was successful in delivering the translation to Prof. Pinkston before his travel.

2.4. *Recollection on my PhD qualifying exam*

It soon became time to schedule my PhD "qualifier" exam and to begin studying for it. After observing the physical and mental stress exhibited by several students who took the exam for a second time, coupled with my still lacking self-confidence, I promised myself that I would take it only once. And that I should pass it by a sufficient margin that Joe would not have to "fight" for my pass in the faculty committee meeting. Much of over five months went into my preparation and success. I sincerely appreciated Joe's support, knowing that my research time suffered. There even came a significant life lesson in this preparation. I learned better to break down complex problems using fundamental and sometimes "simplistic" approaches. My confidence had by now grown! More on this in the following.

2.5. *Recollection on helping to prepare Joe's presentation to the Soviet Academy of Sciences*

In late 1974, after I defended my PhD thesis, we senior graduate students were asked to prepare presentation material for Joe's historic address to the Soviet Academy of Sciences. We first gathered research results and made drawings suitable for publication-quality photographs, slides, and transparencies. VU's Medical Illustration department would typically be our preferred vendor for making such.

Joe's presentations typically contained many figures. We were probably about two-thirds done when Medical Illustration told us that we passed its completion deadline in time for Joe's travel. About two days before Joe was to leave for his trip, we finished the additional drawings. Fortunately, my colleagues, Stephen Lee, and Jerry Cole possessed very nice cameras, and Jerry knew how to develop film.

The three of us set up "industrial lighting" in a laboratory with the appropriate floor space to lay out the drawings for photography. During the first evening, the photographs were taken, and the film was developed in the middle of the night. To our horror, the developed film was blank owing to the failed camera shutter! After some sleep, we met again. Unfortunately, Stephen was unavailable. So, Jerry and I set up the lighting, staged the figures, and took photographs using the second camera. Over this second night, Jerry developed the film and produced prints. The prints were useable but had a light gray background. There was no time to redo the process as Joe's flight was to leave Nashville this morning.

After the last prints finished drying, Abu-Zeid Ahmad volunteered to drive me to the airport. I rushed to Joe's gate. Joe was running more quickly towards me! My high-school football coach would have been proud (I was a lineman) of my handoff of the photos to Joe. Using the skills of his running back days at Mississippi College, he took the handoff and raced back to his gate. As I started to return to the departure area, I saw a stationary plane on the tarmac. I still wonder if that was Joe's plane and if he had it held! A few days later, we received a postcard from Joe in New York City requesting us to have the drawings professionally rephotographed! However, Joe did not forget to thank us during his travels. I received and still treasure his postcards from Russia (see Fig.1).

3. On Joe Sharing with Students and Colleagues

Joe graciously shared his joy of accomplishment and thanked those who helped him. After Joe was awarded the Jesse Beams Gold Medal for Outstanding Research at the Southeastern Section of the American Physical Society meeting in 1975, I received a thank-you card that I also treasure (Fig. 2).

4. Working for Joe at ORNL

Soon after my PhD defense, Joe called me aside and asked me to take a research associate position with him but to live in Oak Ridge, TN. He managed to have the position created jointly as an ORNL Consultant with the Metals and Ceramics and the Physics Divisions. I was deeply grateful for this mentoring as new scientific opportunities, collaborations, and friendships were realized in further gamma spectroscopy, Coulomb excitation, and Coulomb nuclear interference studies over the next two years.

Visiting Gene Eichler and Noah Johnson over lunch at the ORNL cafeteria was always educational and fun! I mentioned an intriguing result in our recent ^{164}Er(α, α') reaction study.[5] They suggested using a heavy ion to excite the target, e.g., the (^{16}O, ^{16}O'γ) Coulomb-excitation reaction. Several experiments investigating the high-spin structure of ^{164}Er were very soon planned and executed along with other studies.[16–18] Gene and

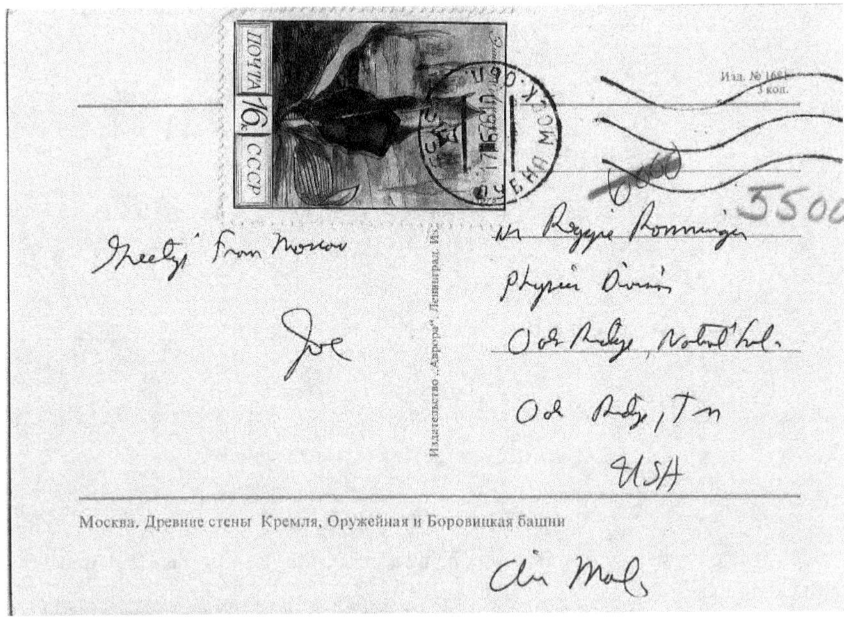

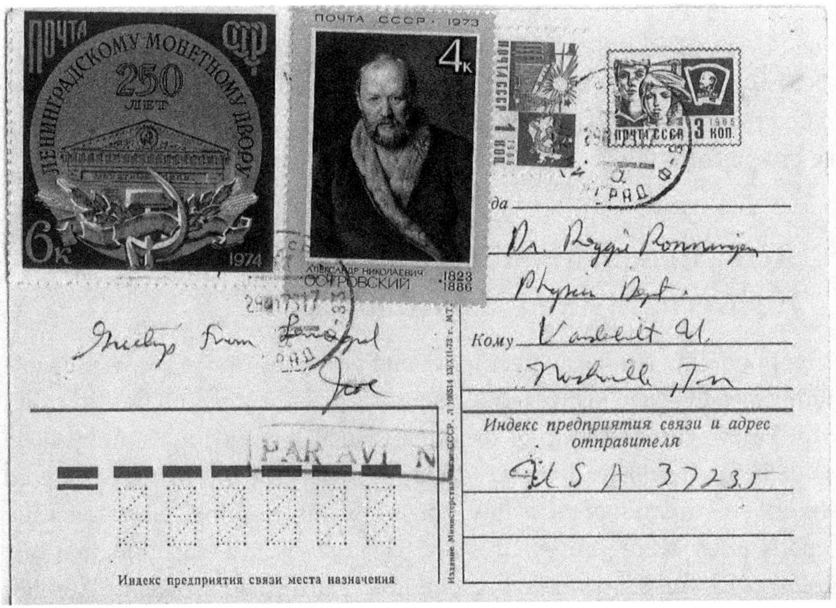

Fig. 1.   Postcards from Joe when in Russia.

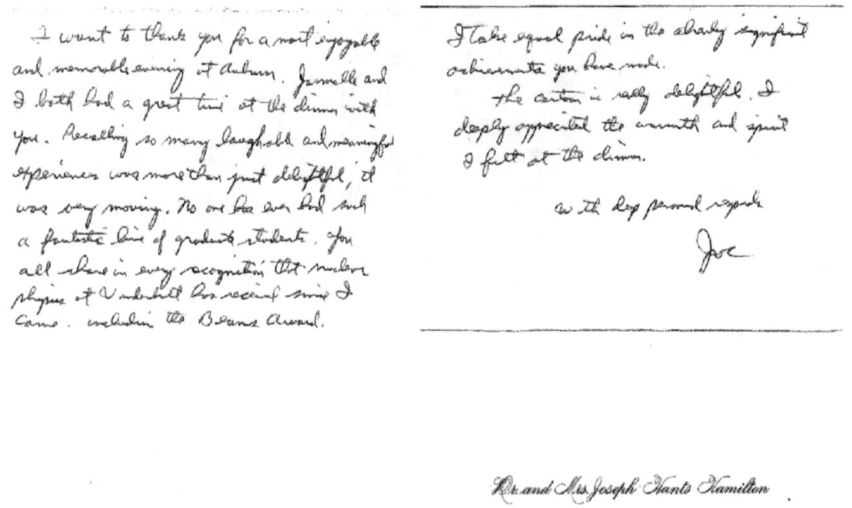

I want to thank you for a most enjoyable and memorable evening at Auburn. Jannelle and I both had a great time at the dinner with you. Recalling so many laughs and meaningful experiences was more than just delightful, it was very moving. No one has ever had such a fantastic line of graduate students. You all share in every recognition that nuclear physics at Vanderbilt has received since I came, including the Beams Award.

I take equal pride in the already significant achievements you have made.

The cartoon is really delightful. I deeply appreciated the warmth and spirit I felt at the dinner.

With deep personal regards

Joe

Dr. and Mrs. Joseph Hants Hamilton

Fig. 2. A thank-you card from Joe after being awarded the Jesse Beams Gold Medal for Outstanding Research.

Noah even called me from LBNL one evening and asked if a thick target of ^{164}Er was available and, if so, would I hand-carry it to LBNL. Joe permitted me to do so, and beautiful results were obtained. In the fall of 1976, we did perform the (^{16}O,^{16}O'γ) Coulomb excitation experiment.

5. Reflection on My Year at the Max-Planck-Institut für Kernphysik in Heidelberg

After pondering my next career move and reflecting on the knowledgeable European and Asian visitors to Joe's group, I took a research position with Prof. Peter Wurm's group at the Max-Planck-Institut für Kernphysik in Heidelberg, Germany. Joe and Ramayya's visitor from there, Dr. Bernd Martin, was instrumental in my obtaining this position. New scientific opportunities were presented with research at this institute and the Gesellschaft für Schwerionenforschung in Darmstadt, Germany. Another opportunity arranged by Joe was for me to present at an international conference in Dresden (then in East Germany). This opportunity provided a fantastic scientific and also profoundly personal experience.

At the Max-Planck-Institut, I met Prof. Jerry Nolen, who was on sabbatical from Michigan State University. He suggested I apply for a position at the MSU Cyclotron Laboratory, for we had common research interests. He even had a graduate student from UWRF, Robert Melin. I learned later that the same doctor delivered Bob and me at the hospital in my small hometown of Frederic, WI, five years and a day apart!

6. Reflection on My Michigan State University Years

During my first several years at MSU's Cyclotron Laboratory, I worked with graduate students Paul Deason, Joe Finck, and Bob Melin to help them take and analyze data for their theses and help with writing papers. The common focus was to elicit nuclear shape data using proton inelastic scattering. Thus, this enabled me to expand on my early work with Joe. These publications[19–23] were instrumental to our analyses and interpretations of the data, later confirmed by a Japanese group.[24] I was proud to present a summary of our findings to Joe and others, along with an independent confirmation at a symposium at VU honoring Joe.

7. Recollection on Joe and Sports Cars

As a graduate student and soon after I joined Joe's research group, I went to ORNL for my first visit to attend a UNISOR collaboration meeting. Afterward, Joe offered me a ride back to VU in his Fiat 124 Spider. I was so impressed with the ride that I soon purchased a used Spider. Now, he did warn me of high maintenance costs. And I soon discovered that the one I bought had been abused and it had been previously repaired after an accident. Over the years, I had to have the transmission rebuilt, the motor block replaced, and the frame rewelded. Jerry Cole still reminds me about the part he machined to replace the cooling fan's failed electromagnetic clutch system! Overall, I learned that Fiats are fun to drive (I also have owned a 124 Sedan and an X1/9) but only when running well!

From late 1985 to 2015, I owned an Alfa Romeo Spider, which was better engineered than the Fiat. I later learned from Joe that he also owned one! When I asked him why he chose the Alfa, he told me it was a sports car with a trunk big enough to hold his golfing gear! Over the years, I have enjoyed discussing cars with Joe. He even allowed me to drive his Chrysler Prowler (Fig. 3) during one of my visits to VU.

Several years ago, I took friends to tour the recording studio, Paisley Park, established by the recording artist Prince. During the tour, we viewed Prince's cars, one of which was a purple Plymouth Prowler. I sent a picture to Joe, and he responded with the history that all first-delivered Prowlers were purple.

I still carry on the sports car tradition with my delightfully designed and well-built 2019 Fiat 124 Spider. In the future, I plan to have Joe and Jannelle take it out for a top-down spin.

Do all of you know that Joe went to high school with and was several years senior to a famous rock and roll artist? Yes, Jerry Lee Lewis. I was entertained by Joe's story about meeting Jerry Lee after a concert. Jerry Lee was amazed at how young Joe looked and asked Joe, "Did I flunk that many grades?"

Fig. 3.   Joe and his Chrysler Prowler.

8. Visits to Joe

When visiting Nashville, a visit to Joe is an itinerary must. Here is a photo of Joe with Jerry and Carolann Cole in his office during one such visit (Fig. 4).

Joe's final Colloquium at VU was an event not to be missed. Neither was the prior-event social put on by the Physics and Astronomy Department staff nor the post-event dinner hosted by Joe and Jannelle. These were great times with great friends (Fig. 5).

9. Future Endeavors

In the fall of 1976, as Gene Eichler and Noah Johnson suggested, we performed the $(^{16}O, {}^{16}O'\gamma)$ Coulomb excitation experiment on ^{164}Er. This work is still unpublished for several reasons: The student assigned to

Fig. 4. Carolann Cole, Joe Hamilton, and Jerry Cole in Joe's office at Vanderbilt University.

Fig. 5.   Joe and Jannelle with former students and spouses after Joe's retirement collo-quium at Vanderbilt University.

perform the analysis left the university for new endeavors, and the photon detector was not correctly calibrated for efficiency. However, I still main-tain paper copies of the data and believe the data are still publishable.

10. Reflections During Retirement

Now retired, I can reflect on my attempt to seek a career in health physics after college but instead obtain a PhD and have a career in nuclear physics. Did I meander too much? No, I "completed the circle." At MSU in 1982, Henry Blosser assigned me the laboratory radiation safety officer position. Over the years, I grew in this position, which allowed me to contribute to establishing NSCL and FRIB.

As promised, I now elaborate on breaking down complex problems using fundamental and sometimes "simplistic" approaches. For example, at MSU, we extended[21] George Bertsch's beautiful and straightforward explanation[25] describing the evolution of quadrupole and hexadecapole nuclear shapes between closed shells to the next-higher hexacontatetrapole order. Another of my most satisfying contributions was calculating personnel radiation doses from uncontrolled beam losses in a heavy-ion linear accelerator after the beam shuts off. This was accomplished by first using semi-phenomenological equations and assumptions found in the book by A. H. Sullivan.[26] The result was confirmed using a Monte Carlo-based radiation transport code and extended to heavy ion beams.[27]

In both examples, a simple approach led to something useful and easy to communicate with others. In a way, the epigram in Sullivan's book *A Little Inaccuracy Sometimes Saves Tons of Explanation*, served me well! However, these efforts were based on my studies to pass that PhD qualifying examination. Thus, I owe much of my professional and other achievements in life to the attributes Joe passed down. Reviewing the list of my friends and collaborators in my scientific and personal life, I know I have been blessed by having Joe, Ramayya, and many other outstanding scientists.

References

1. R. M. Ronningen, J. H. Hamilton, A. V. Ramayya, L. Varnell, G. Garcia-Bermudez, J. Lange, W. Lourens, L. L. Riedinger, R. L. Robinson, P. H. Stelson, and J. L. C. Ford, Jr., Reduced Transition Probabilities of Vibrational States in $^{154\text{-}160}$Gd and $^{176\text{-}180}$Hf, *Phys. Rev.* **C15**, 1671 (1977).
2. R. M. Ronningen, J. H. Hamilton, L. Varnell, J. Lange, A. V. Ramayya, G. Garcia-Bermudez, W. Lourens, L. L. Riedinger, F. K. McGowan, P. H. Stelson, R. L. Robinson, and J. L. C. Ford Jr., E2 and E4 Reduced Matrix Elements of ^{156,158,160}Gd and ^{176,178,180}Hf, *Phys. Rev.* **C16**, 2208 (1977).
3. R. M. Ronningen, J. H. Hamilton, and A. V. Ramayya, B(E2;0g+IK = 2+0) of the 1276.7 keV Transition in ^{178}Hf, *Phys. Rev.* **C65**, 047304 (2002).
4. R. M. Ronningen, R. B. Piercey, J. H. Hamilton, C. F. Maguire, A. V. Ramayya, H. Kawakami, B. van Nooijen, R. S. Grantham, W. K Dagenhart, and L. L. Riedinger, Coulomb Excitation Measurements of Reduced E2 and E4 Transition Matrix Elements in ^{156,158}Dy, ^{162,164}Er and ^{168}Yb, *Phys. Rev.* **C16**, 2218 (1977).

5. R. M. Ronningen, R. S. Grantham, J. H. Hamilton, R. B. Piercey, A. V. Ramayya, B. van Nooijen, H. Kawakami, W. Lourens, R. S. Lee, W. K. Dagenhart, and L. L. Riedinger, Reduced Transition Probabilities of Vibrational States in ^{156,158}Dy, ^{162,164}Er and ^{168}Yb, *Phys. Rev.* **C26**, 97 (1982).

6. A. Ahmad, G. Bomar, H. Crowell, J. H. Hamilton, H. Kawakami, C. F. Maguire, W. G. Nettles, R. B. Piercey, A. V. Ramayya, R. Soundranayagam, R. M. Ronningen, O. Scholten, and P. H. Stelson, Coulomb Excitation of ^{144,146,148,150}Nd, *Phys. Rev.* **C37**, 1836 (1988).

7. W. G. Nettles, A. V. Ramayya, J. H. Hamilton, R. Soundranayagam, H. Yamada, T. Humanic, J. X. Saladin, A. Hussein, R. B. Piercey, and R. M. Ronningen, Coulomb-Nuclear Interference Measurements of Hexadecapole Deformations in ^{168}Yb and ^{178,180}Hf, *J. Phys.* **G14**, L223 (1988).

8. R. M. Ronningen, R. B. Piercey, A. V. Ramayya, J. H. Hamilton, S. Raman, P. H. Stelson, and W. K. Dagenhart, Coulomb Excitation of 2+ and 3- States in ^{192,194}Pt, *Phys. Rev.* **C16**, 571 (1977).

9. R. M. Ronningen, F. T. Baker, Alan Scott, T. H. Kruse, R. Suchannek, W. Savin, and J. H. Hamilton, Evidence for A Nonrotational Interpretation of <0+‖M(E4)‖4+> in ^{180}Hf, *Phys. Rev. Lett.* **40**, 364 (1978).

10. F. Todd Baker, Alan Scott, R. M. Ronningen, T. H. Kruse, R. Suchannek, W. Savin, and J. H. Hamilton, Sensitivity of (α,α') Cross Sections to Excited-State Quadrupole Moments, *Phys. Lett.* **70B**, 167 (1977).

11. F. Todd Baker, Alan Scott, R. M. Ronningen, T. H. Kruse, R. Suchannek, and W. Savin, ^{192}Pt(α,α') Reaction at E=24 MeV, *Phys. Rev.* **C17**, 1559 (1978).

12. J. H. Hamilton, A. V. Ramayya, W. T. Pinkston, R. M. Ronningen, G. Garcia-Bermudez, H. K. Carter, R. L. Robinson, H. J. Kim, and R. O. Sayer, Evidence for Coexistence of Spherical and Deformed Shapes in ^{72}Se, *Phys. Rev. Lett.* **32**, 239 (1974).

13. A. V. Ramayya, R. M. Ronningen, J. H. Hamilton, W. T. Pinkston, G. Garcia-Bermudez, R. L. Robinson, H. J. Kim, H. K. Carter, and W. E. Collins, Mean Life and Collective Effects of the 937 keV 0+ State in ^{72}Se: Evidence for Nuclear Coexistence, *Phys. Rev.* **C12**, 1360 (1975).

14. R. M. Ronningen, A. V. Ramayya, J. H. Hamilton, W. Lourens, J. Lange, H. K. Carter, and R. O. Sayer, Mean Life of the 854 keV 0+ State in ^{74}Se and the Coexistence Model, *Nucl. Phys.* **A261**, 439 (1976).

15. J. H. Hamilton, H. L. Crowell, R. L. Robinson, A. V. Ramayya, W. E. Collins, R. M. Ronningen, V. Maruhn-Rezwani, J. A. Maruhn, N. C. Singhal, H. J. Kim, R. O. Sayer, T. Magee, and L. C. Whitlock, Lifetime Measurements to

Test the Coexistence of Spherical and Deformed Shapes in ^{72}Se, *Phys. Rev. Lett.* **36**, 340 (1976).

16. Noah R. Johnson, D. Cline, S. W. Yates, F. S. Stephens, L. L. Riedinger, and R. M. Ronningen, Multiple Band Crossings in ^{164}Er, *Phys. Rev. Lett.* **40**, 151 (1978).

17. S. W. Yates, I. Y. Lee, N. R. Johnson, E. Eichler, L. L. Riedinger, M. W. Guidry, A. C. Kahler, R. S. Simon, P. A. Butler, P. Colombani, F. S. Stephens, R. M. Diamond, R. M. Ronningen, R. D. Hichwa, J. H. Hamilton, and E. L. Robinson, High-spin Properties of ^{164}Er in the Multiple Band Crossing Region, *Phys. Rev.* **C21**, 2366 (1980).

18. J. H. Hamilton, A. V. Ramayya, R. M. Ronningen, R. O. Sayer, H. Yamada, C. F. Maguire, P. Colombani, D. Ward, R. M. Diamond, F. S. Stephens, I. Y. Lee, P. A. Butler, and D. Habs, Isomeric Trapping Following Coulomb Excitation of High Spin States in ^{178}Hf, *Phys. Lett.* **112B**, 327 (1982).

19. P. T. Deason, C. H. King, R. M. Ronningen, T. L. Khoo, F. M. Bernthal, and J. A. Nolen Jr., ^{194,196,198}Pt(p,p') Reactions at 35 MeV, *Phys. Rev.* **C23**, 1414 (1981).

20. C. H. King, J. E. Finck, G. M. Crawley, J. A. Nolen Jr., and R. M. Ronningen, Multipole Moments of ^{154}Sm, 176Yb, ^{232}Th, and ^{238}U from Proton Inelastic Scattering, *Phys. Rev.* **C20**, 2084 (1979).

21. R. M. Ronningen, R. C. Melin, J. A. Nolen Jr., G. M. Crawley, and C. E. Bemis Jr., Higher-Order Deformations of ^{232}Th and ^{234,236,238}U, *Phys. Rev. Lett.* **47**, 635 (1981).

22. R. M. Ronningen, G. M. Crawley, N. Anantaraman, S. M. Banks, B. M. Spicer, G. G. Shute, V. C. Officer, J. M. R. Wastell, D. W. Devins, and D. L. Friesel, Multipole Moments of ^{154}Sm and ^{166}Er by Inelastic Scattering of 134 MeV Protons, *Phys. Rev.* **C28**, 123 (1983).

23. B. G. Lay, S. M. Banks, B. M. Spicer, G. G. Shute, V. C. Officer, R. M. Ronningen, G. M. Crawley, N. Anantaraman, and R. P. DeVito, Multipole Moments of ^{176}Yb and ^{182}W Nuclei from Inelastic Scattering of 134 MeV Protons, *Phys. Rev.* **C32**, 440 (1985).

24. T. Ichihara, H. Sakaguchi, M. Nakamura, T. Noro, F. Ohtani, H. Sakamoto, H. Ogawa, M. Yosoi, M. Ieiri, N. Isshiki, and S. Kobayashi, T. Ichihara, H. Sakaguchi, M. Nakamura, T. Noro, F. Ohtani, H. Sakamato, H. Ogawa, M. Yosoi, M. Ieiri, N. Isshiki, and S. Kobayashi, Multipole Moments of Er^{166}, Er^{168}, Yb174, and Yb^{176} from 65 MeV Polarized Proton Inelastic Scattering and Density Dependence of the Effective Interaction, *Phys. Rev.* **C29**, 1228 (1984).

25. G. F. Bertsch, Remark on Y4 Moments, *Phys. Lett.* **B26**, 130–131 (1968).
26. A. H. Sullivan, *A Guide to Radiation and Radioactivity Near High Energy Particle Accelerators*, Nuclear Technology Publishing, Ashford, Kent, England (1992).
27. R. M. Ronningen, G. Bollen, and I. Remec, Estimated Limits on Uncontrolled Beam Losses of Heavy Ions for Allowing Hands-On Maintenance at an Exotic Beam Facility Linac, *Nucl. Tech.* **168**, 670–675 (2009); Science Technical Report 116, AT&T Bell Laboratories, Murray Hill, NJ (1984).

My Professor: Joseph H. Hamilton

Rhett Wiseman

I played baseball at Vanderbilt from 2013 to 2015 and was drafted in the 3rd round of the MLB draft in 2015. I went back and finished my degree in 2018 while playing professionally. I now own a real estate development company based out of Nashville. Below I would like to tell you about taking physics classes from Professor Joseph Hamilton, about him always being in the crowd cheering us on at baseball matches, and the huge impact he has had on me.

The author waving to Joe Hamilton who is seated in his usual place behind the dugout. Team member Ben Bowden is in the foreground. Photo taken by JHH.

It isn't often that the most distinguished professor at a top university in the world is also one of the most distinguished sports fans on campus. Joseph Hamilton was THAT professor his entire teaching career. As a sophomore on the 2014 National Championship Vanderbilt Commodores Baseball team, Professor Hamilton was a staple in the crowd. A face we all loved to see cheering us on no matter what the scoreboard looked like. That season, it was easy to be a fan of ours, but that was not always the case.

Through his incredible career as a highly popular and insanely successful nuclear physicist, Professor Hamilton was better known among laymen like me as the little physics professor who was willing to do ANYTHING to help his students succeed. Athletes on campus flocked to Joe not because his classes were easy (in reality they were anything but!) but rather because of his genuine care for other people, his ability to TEACH at a level far below his own personal understanding of the subjects discussed, and his love of sports.

As an athlete at a top university, there were always whispers of the classes you should avoid and the classes you should take. I can remember sitting in the baseball locker room back in 2013 with my fellow freshman teammates and having the upperclassman tell us that we MUST take as many classes with Professor Hamilton as possible. They discussed how no matter how hard the material of the class was, Professor Hamilton would spend time before class, after class, or in any free time he had, to help us understand, pass, and LEARN in his class. The upperclassmen discussed how this Professor not only cared about his future Rhodes Scholar students, but also the baseball players who were grinding through Physics 101 in hopes of passing.

I listened to my older teammates and signed up for physics that following fall. I quickly found out that everything I had heard was true! When I first walked into the classroom in the fall of 2014, I was SURROUNDED by athletes. I couldn't tell if I was in an athletic department meeting, or Physics 101! At 9 am when class began, a little older man appeared from an office at the front of the room wearing a brand-new football jersey, and an eruption from one corner of the room bellowed throughout the auditorium. The football-player corner. The class began with Professor Hamilton introducing himself and then discussing the accomplishments that every sports team had achieved that past weekend.

Professor Hamilton was always able to grab the attention of all the athletes immediately and able to hold their attention for the entire class. Not an easy feat!!

In 2014, the Vanderbilt Commodores baseball team made history. Every step of the way, Professor Hamilton was there with a welcoming smile. And then again in 2015 when we went back to the College Baseball World Series, he was there again. I'll never forget how he made the trips to Omaha Nebraska, and waited to talk to his students on the team after every game. He would be there waiting for us as we loaded up the bus and was there when we got off. Win or lose. During the games, he would sit right behind the dugout with his camera, and take so much joy in getting great candid photos of us while we messed around in the dugout, and on the field. It's one thing to come to sporting events on campus that are just a few hundred yards away from your classroom. It's another thing to travel 800 miles, spend thousands of dollars, and witness school history for the first time. Those things don't go unnoticed, and just another reason why so many of us cherish Professor Hamilton as much as we do.

The more time I spent in the Professor's classes, the more I learned, and the more we talked during and after class. Office hours turned into not only learning about physics, but also discussing baseball, sports, history, and life. I cherished the time we spent together talking about so many different things! After I had left Vanderbilt following the 2015 MLB draft, I needed to take on an independent study project to graduate, and who better to take that with, than the Professor himself! With the help of Joe, I was able to take a DEEP DIVE into the history of the Manhattan Project, the development of the Atomic Bomb, and the development of the Hydrogen bomb thereafter. Learning about not only the history but also the science and details of what made those projects so incredible and important to human history.

Prior to Professor Hamilton's class, I never had an interest in nuclear physics or nuclear science. So many athletes who dedicated most of their time to their sport would often find themselves learning more in Joe's class than anywhere else on campus. Why was that? Was it because the material was better than other classes? No. Was it because Professor Hamilton was the most distinguished on campus? No. It was his ability to connect with students and influence them in a way that other Professors could not. It was a genuine care for all of his students no matter what their capability was in the classroom, that led so many people to respect him and listen to his words, whether they loved physics or didn't. Joseph Hamilton has always been able to connect with so many different people and is so passionate about other human beings' success. Joe has been able to keep valuable relationships and strong communications with so many of his students, that the student/professor relationship has blossomed into so many real friendships. I am happy and proud to call myself one of those students, and more importantly, one of those friends.

Joseph H. Hamilton's Final Physics Class

Hunter Owen

*Vanderbilt Baseball Player and Student,
Nashville, TN 37235, USA*

I was lucky enough to be a student in Dr. Hamilton's final physics class before his retirement. It was a truly extraordinary experience which I will never forget.

Physics class with Professor Hamilton was truly one of the most engaging and educational courses I have taken at Vanderbilt. This course was an introduction to physics and provided the foundational elements of physics. One of the more entertaining experiments that Professor Hamilton demonstrated in class was when he laid on a bed of nails. At the beginning of class, as everybody was walking in and getting their seats, the bed of nails was covered with a cloth so none of us could see what it was. However, as he pulled the cloth off of the bed of nails everybody marveled in shock at what was going to happen next. I'm sure Professor Hamilton demonstrated this experiment a hundred times to his class. Except in this certain experiment I forgot to mention that he was also 90 years old. As he laid on the bed of nails, his assistant put the other set of nails on top of him so he would not be impaled. This experiment was truly an amazing site to see. This was not the only experiment that he performed in class. Every day he would come to class with a different experiment that would keep the class in shock. Even though Professor Hamilton performed all these amazing experiments in class, he was also a super helpful and kind professor. It truly felt as if he would stop at nothing to help you understand the appreciation of physics and how it all works. I always admired his love for what he was teaching and also his enthusiasm every single day. Considering that he was part of a group that discovered an element on the periodic table, I can probably tell you how intelligent he is. However, the thing that separated Professor Hamilton is that while he had all of this intelligence he truly wanted the next generation of students to obtain the same knowledge that he had throughout his career. Being a part of his last class at Vanderbilt was something pretty cool. You can definitely see how much he is loved by his peers and the people he interacts with every single day. On the last day of class, we all gave him a standing ovation and I felt

Fig. 1.   The author with Joe together at another Vanderbilt win in February 2023.

so honored to be a part of the legacy that he had built at Vanderbilt throughout the years.

Postscript: Joe Hamilton writes: *Not only was Hunter a good student and often helped me do class demonstrations, he is now one of the best pitchers in college baseball. I watched him pitch against the U. of Mississippi on March 17th, 2023 at Hawkins Field in Nashville. U. of M. are last year's national champion and currently ranked number three in the nation. They are also the number one team in hitting and runs in the Southeastern Conference which has four teams in the top five nationally. Hunter allowed only two hits and struck out 11 including their top two hitters in the top of the last inning to win 8 to 0!! Congratulations Hunter!*

Mentoring: A Lifelong Activity

Joseph H. Hamilton

*Department of Physics and Astronomy, Vanderbilt University,
Nashville, TN 37235, USA*

The following remarks were made in my talk summarizing my lifelong mentoring activities when I received the Mentoring Award given by the Nuclear Physics Division of APS at its fall meeting in 2016.

First, I want to express my deep thanks to my former students and colleagues who nominated me for this award. There is a real sense in which mentoring a student or postdoctoral fellow or even colleague is like having a child; they are part of you, and like with your children, good mentoring becomes a lifelong activity. My graduate students and I are like a close-knit family. We share in each other's lives, often continuing to do research together, and take pride in each other's successes. I have been extraordinarily blessed to have had over 60 PhD students, 30 MS, 10 senior honors, 8 students Research Experience for Undergraduate, and over 100 research associates with whom I have shared life, as well as over 10,000 undergraduates whom I have taught physics.

Isaac Newton once said, "If I have seen far it is because I have stood on the shoulders of giants." Indeed much of my success in research is because of the marvelous work of my students and colleagues. Many of them have gone on to impressive careers in research and teaching. I am very proud of the accomplishments of all of them. Many have become faculty members, 17 became department chairs, one dean, one provost, several vice presidents for research, and numbers went to national labs including one who became Deputy Director of ORNL, several at DOE headquarters, and two vice presidents of large companies. Many have won national awards, and 6 Southeastern Section APS awards: 3 for teaching, 1 for research, and 2 for service, one former student won two $10,000 Awards from his company and two Industrial Research IR 100 awards for the most outstanding technical developments for the year and one former student is currently the PI on two multi-BILLION dollar contracts from the federal government, one former student won a DOE Award for Mercury cleanup in ORNL and 6 of my REU students won DNP awards to present their research at our fall conferences, two here in Vancouver. In this extended family, I have 11 grandchildren who did their undergraduate work with one of my PhD graduates and then came and did their PhD with me and two great-grandchildren who took their undergraduate work with one of my grandchildren and then came for the PhD with me along with another 12 grandchildren who took their PhD under one of my former research associates at another university where they used our large data set for their thesis work.

I would like to say thanks to all these students and collaborators and share with you pictures of some of them. I would first like to thank those who mentored me in teaching and research, my PhD thesis director, Larry Langer, Guy Forman who taught me the importance of doing demonstrations in teaching beginning physics, Robert Lagemann who funded my first two research associates and introduced me to teaching non-science majors and Ernie Jones who helped me to see the importance of office hours for students and Wendell Holliday who nominated me for my first research award. From Prof. Slack I secured funds to endow a lecture series that brought many Nobel Laureates to Vanderbilt. Ben Mottleson was one of them.

Prof. A. V. Ramayya came to me as a research associate in 1964, became a senior research associate in 1968, and became an assistant professor in 1970, rising to full professor in 1980. We have co-mentored many of these graduate students and research associates and I am deeply indebted to him for all our collaborations.

In 1963, my first group of graduate students gave me a birthday party, four of who started with me in 1960 the "Big Team". They remained close lifelong friends and for the last 10–15 years took annual summer trips together. When I was in Amsterdam from 1962–1963 I would send them long handwritten letters about what they should be doing in research. Unfortunately, I am known for having terrible handwriting. Years later one told me when they got one of my letters they would make three more copies and each would take it home to write out their translation. By combining all four they could decide what was said. Last summer I got a postcard from Appalachia where the four took their trips. They wrote, "we want to thank you again for all you did for us in our thesis work over 50 years ago."

Many of my students played pivotal roles in our development of new facilities for research at Oak Ridge National Lab. The first project was the University Isotope Separator at Oak Ridge, the first isotope separator online to a cyclotron for studies of nuclei far from stability, Ed Zganjar, Billy Brantley, Lee Riedinger, Caroll Bingham, Gene Spjewski, and Ken Carter (who became director of the project) played important roles. This was 1970, and federal funds were very tight. Several of my former students as well as me were about to dry up on the van with research to do on campus. "In these days of austerity in the physical sciences, it's a particular pleasure to be able to point to a project that goes counter to that trend.

UNISOR is remarkable, not only for its sheer existence, but also for the extraordinary measure of cooperation between so many institutions that is represents." — Alvin M. Weinberg, Director of Oak Ridge National Laboratory. At the dedication, June 19, 1972. When I was raising money for the UNISOR project, a good friend of mine was chief of staff for Governor Ellington. In telling him about our desire to build the separator, he invited me to make a proposal to the State Building Commission for funds to match the $90,000 from the university so he helped us double our money. He subsequently told me the following. He said, "Joe, every government agency at the local, state, or federal level gets more requests for funds than they can possibly fund. In general, they fund those projects from people they know and know they can trust that what they have proposed they can do." That was a very important lesson for me. And I hope for the young scientists here.

A second project was the Joint Institute for Heavy Ion Research at ORNL, a cooperation of Vanderbilt, U. Tennessee, and ORNL. The institute played an important role in making the Holifield an international center for nuclear research. Cooperation with Lee Riedinger (UT) and Russell Robinson (ORNL) played a major role in this development. When we wanted funds for a second building for our Joint Institute, I asked our Chancellor whom I knew well to ask then Governor Alexander for such funds. The governor honored the request. "In basic nuclear research, we have two outstanding success stories — the University Isotope Separator at Oak Ridge (UNISOR) and the Joint Institute for Heavy Ion Research" — Edward A. Friedman, Director Office of Energy Research, US Department of Energy, 1980.

I worked to become friends with State Senator Henry from Nashville who was chair of the Senate Appropriations Committee by inviting him to Vanderbilt and Oak Ridge to discuss our projects. When I proposed a new generation recall mass separator for Oak Ridge along with my several former students, Argonne put in a proposal for a current style RMS for ¼ the cost. So, DOE turned down our proposal. I then raised nearly a million dollars from non-DOE money and asked the State of Tennessee for another $500,000. Senator Henry arranged a meeting for me and Lee Riedinger from UT to seek Governor McWorther's approval for the funds. Just as an aside, the governor asked me how much VU was pledging. I said $150,000 and then asked Lee what was UT pledging again $150,000. The Governor said "How is that possible? They don't have two dimes to rub together. I should take that RMS and shove it some place and find out where they are getting the money." After lengthy discussions supported by Senator Henry, the governor turned to his finance director and said, "Are you going to give me a hard time if I approve this." To which the director replied, "The last time I looked you were the Governor." After informing DOE of our nearly 1.5 million dollars of non-DOE funds, I got a call from Dave Henry, DOE's nuclear physics program officer, saying stop, you can have the rest of your 4 million dollars. This is the RMS that one of my research associates J. J. Das and PhD student Tom Ginter helped put online.

A few years later DOE put out a call for proposals to build a large Compton-suppressed Ge detector array. They indicated local support would be important but gave only one month to the labs to get local support. Oak Ridge called me about getting State support. First I got VU and UT to pledge $100,000 each, then I went to see Senator Henry and said we need to get the State to put up $800,000 for this $20,000,000 detector array to get it put at ORNL. After brief discussions, he picked up his phone and called the state finance director and said, "I have Prof. Hamilton in my office and he and his ORNL colleagues need an additional $800,000 to help bring a new 20 million dollar facility to Oak Ridge. He needs a letter signed by the governor by next Monday pledging these funds." Then he said, "So this is okay? All right I am sending Prof. Hamilton to you now to work out the details." I told my wife I was amazed to walk into an office and in 30 minutes walk out with $800,000 in my pocket. We later heard the strong Tennessee support influenced the Secretary of Energy to fund Gammasphere. Senator Henry also got the state to fund half the cost of the total theory wing of the Joint Institute.

There were other ORNL projects. In 1988, a nuclear orientation facility was completed at UNISOR. Six of my graduate students and 3 of my former graduate students and research associates worked to get this facility completed.

Now let me show a few pictures to illustrate the relationships with my former students. This is a dinner with them following my receiving the 1975 Beams Award for Outstanding Research given by the Southeastern Section of APS.

Among the over 10,000 undergraduates I have taught, there have been many athletes. Here are 4 of the men's basketball team that defeated the University of Kentucky for the 2012 SEC championship, one of only two losses for Kentucky that went on to win the NCAA championship. I taught all 14 members of that team. At an earlier game that year I was honored at one of their basketball games and presented with a basketball signed by every member of the team.

In 2014, the VU baseball team won the College World Series in Omaha. I taught 25 of the 28 members of that team and have a signed baseball by them. This is a picture of Rhett Wiseman, the star right fielder, with me at breakfast at Omaha. When it was announced on June 8, 2016, that the new element that I had been a key figure in discovering was to be name tennssine as I had proposed, I received two days later an e-mail from Rhett, who is now playing professional baseball, congratulating me on our discovery and name.

The next figure is one of the 20 years of high school teachers and students I took to ORNL as part of our 3-week summer science project to interest students and teachers in science. They are around the RMS separator at ORNL.

There were a number of international conferences at VU and ORNL that I chaired. At all of the 7 Sanibel conferences, beginning in 1992, many of my former students would come including my first PhD student in 1960. At one they gave me a new birthday party.

Not many physicists can say they have given a physics lecture to 65,000 people at once. I gave six lectures on acceleration, angular momentum, power, action, reaction, etc. at six Tennessee NFL football games. At the first one, during the half, a person came walking down by me and said, "you are the man in the jumbo Tron." I said yes. He said what do you do? I said I teach physics at Vanderbilt. He asked what I teach. I said physics. He said, I thought you were an actor.

There are lots of funny stories but one that stands out in my memory, Lee Riedinger told me many years later this story. He visited several research groups and talked with their students. He asked one of my students how was I to work with. The student responded "he is a good person with whom to work but he is hell to work with two weeks before he is to go to a conference to speak.

Finally, I have been blessed to have had great graduate students, research associates, and collaborators. They have been inspiring to me, enriching my life in many ways. Many of us have formed close friendships that have continued in some cases over 50 years. I love to have get-togethers with them at conferences. It is very much like a family reunion. In closing, I want to thank each one of my students, research associates, and collaborators for everything they have done to make my life wonderful.

Mentoring: A Lifelong Journey

Joseph H. Hamilton
Vanderbilt University
DNP Vancouver, October 2016
(Powerpoint presentation slides)

- Over 60 PhD students
- 30MS (12 stayed on for the PhD)
- 10 senior honors thesis
- 8 REU students
- Over 100 research associates
- Over 10,000 undergraduate students taught

From the
WHOLE
TEAM

The Big Team

"In these days of austerity in the physical sciences, it's a particular pleasure to be able to point to a project that goes counter to that trend. UNISOR is remarkable, not only for its sheer existence, but also for the extraordinary measure of cooperation between so many institutions that it represents." — Alvin M. Weinberg, Director of Oak Ridge National Laboratory. At the dedication, June 19, 1972.

Joint Institute
for Heavy Ion
Research

"In basic nuclear research, we have two outstanding success stories — the University Isotope Separator at Oak Ridge (UNISOR) and the Joint Institute for Heavy Ion Research" — Edward A. Friedman, Director Office of Energy Research, U. S. Department of Energy, 1980.

UNISOR NOF
JUNE 29, 1988

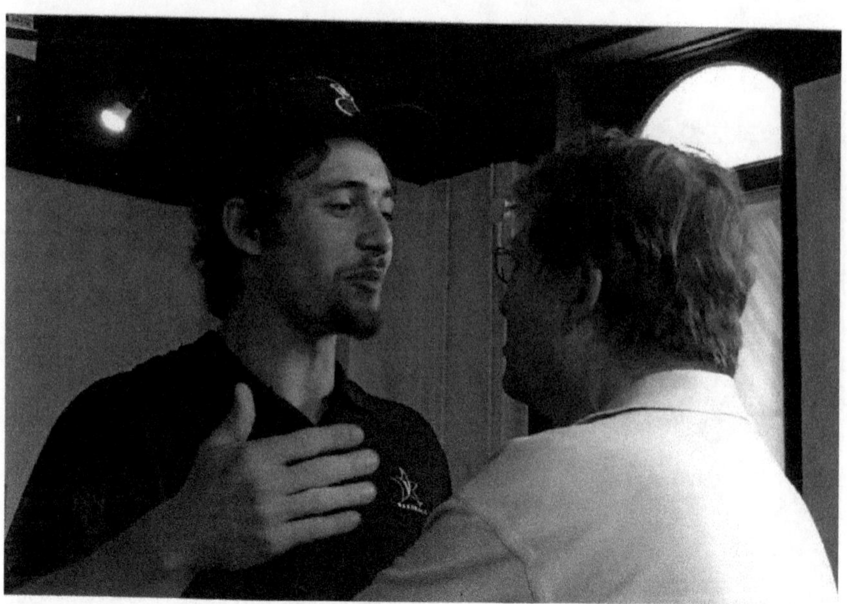

Special thank you to each one of my students, research associates, and collaborators.

Thank you for listening.

Remembering Edward Zganjar (7/31/1938–2/8/2024)

Jannelle and Joe Hamilton, Ed and Jo Zganjar, November 2012.

We were very deeply saddened by the passing of our long-time friend and research collaborator Edward Zganjar who died on February 8, 2024. Ed was a retired Alumni Professor and Professor of Physics at Louisiana State University and an eminent researcher in experimental nuclear physics.

Ed was born July 31, 1938, in Minnesota, grew up there, completed college in 1960, and the same year married Josephine Charmoli, his wife of 63 years. He earned his MS and PhD degrees in nuclear physics from Vanderbilt University in 1963 and 1966 under the direction of Professor Joseph Hamilton. He joined the faculty of LSU in Baton Rouge in 1966 as Assistant Professor, was promoted to Associate Professor in 1970 and full Professor in 1975, and subsequently given the named chair of Alumni Professor. He chaired the Department of Physics and Astronomy from 1982 to 1985 and was Associate Vice Chancellor for Research and Development from 1991 to 1994.

Ed's research interests were in nuclear structure and nuclear astrophysics. He was awarded continuous grant funding from the US Department of Energy from 1970 to 2017. He played a leading role in the development of the university facility (UNISOR/UNIRIB) at Oak Ridge National Laboratory, often elected chair of its Executive Committee, and subsequently in the new TRIUMF facility in Canada. His study of electric monopole transitions played an important role in the discovery of nuclear shape coexistence in neutron-deficient mercury nuclei that helped change the paradigm that every nucleus has one fixed shape. His research also helped provide an understanding of the astrophysical processes that shaped our early universe. In 1982, he was elected a Fellow of the American Physical Society (APS). He received the Francis G. Slack Award for outstanding service from the Southeastern Section of the APS and was inducted into the LSU College of Science Hall of Distinction in 2015.

Perhaps his forte was designing and building equipment which made experiments at HHIRF, HRIBF, TRIUMF, ANL, and FRIB possible. He designed and with his own hands (and the support of the LSU machine shop) built over 30 individual major experimental devices, many of which continue to contribute to leading experimental programs today. His construction creativity extended to his love of woodworking, including making a unique cradle for each of his four children, a magnificent treehouse in a giant oak tree, and a beautiful dining room table that converted into a billiard table for family competitions. He sent Joe a wood carving of all the different nuclear shapes observed in their experiments.

Ed will be missed by his many friends who so enjoyed his outgoing nature, his warm friendship, his humility, and his love of life. The Louisiana expression "let the good times roll" characterized him. From his graduate school days on throughout his life, Ed and Joe developed a relationship almost like son and father. Ed's long-time colleagues (Ken and Lee) were like brothers. Ed was Jeff's primary mentor at LSU and helped him grow an experimental nuclear physics group that now includes three tenured faculty members and twelve graduate students. We four are deeply grateful for all his contributions to our lives.

Laissez les bon temps rouler.
Joseph Hamilton, Ken Carter, Lee Riedinger, and Jeff Blackmon

Lee Riedinger, Ken Carter, Gene Spejewski, and Ed Zganjar on occasion of the 50-year anniversary of the first UNISOR on-line experiment; October 19, 2022.

Joe Hamilton: His Impact on Global Scientific Outreach and International Collaboration

Witold Nazarewicz

*Department of Physics and Astronomy and FRIB Laboratory,
Michigan State University, East Lansing, Michigan 48824, USA*

Remarks presented at the global Zoom celebratory event, discussing in particular, Joe Hamilton's extraordinary impact on global scientific outreach and international collaboration.

1. Introduction

Joe Hamilton has been a champion of international collaborations in nuclear physics. He was one of the early American nuclear physicists to work with Soviet, East German, and Chinese physicists.

2. International Collaborations

Joe has been a leader of major international collaborations with scientists in the People's Republic of China and Russia for 25 years and in France, Germany, India, Slovakia, and Romania.

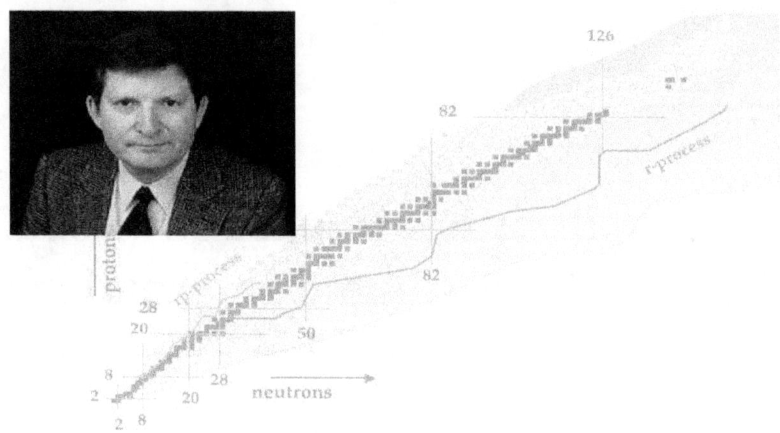

VOLUME 77, NUMBER 1   PHYSICAL REVIEW LETTERS   1 JULY 1996

New Spontaneous Fission Mode for ^{252}Cf: Indication of Hyperdeformed ^{144,145,146}Ba at Scission

G. M. Ter-Akopian,[1,2] J. H. Hamilton,[2] Yu. Ts. Oganessian,[1] A. V. Daniel,[1,2] J. Kormicki,[2,*] A. V. Ramayya,[2] G. S. Popeko,[1] B. R. S. Babu,[2] Q.-H. Lu,[2] K. Butler-Moore,[2,†] W.-C. Ma,[3] S. Ćwiok,[4,5] W. Nazarewicz,[6,7,8] J. K. Deng,[2,‡] D. Shi,[2] J. Kliman,[9] M. Morhac,[9] J. D. Cole,[10] R. Aryaeinejad,[10] N. R. Johnson,[7] I. Y. Lee,[7,§] F. K. McGowan[7] and J. X. Saladin[5]

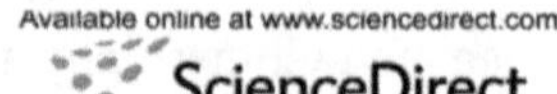
Available online at www.sciencedirect.com

ScienceDirect

NUCLEAR PHYSICS A

ELSEVIER   Nuclear Physics A 834 (2010) 28c–31c

www.elsevier.com/locate/nuclphysa

Super deformation to maximum triaxiality in $A = 100 - 112$; superdeformation, chiral bands and wobbling motion

J.H.Hamilton [a], S.J.Zhu [b], Y.X.Luo [a c], A.V.Ramayya [a], S.Frauendorf [d], J.O.Rasmussen [c], J.K.Hwang [a], S.H.Liu [a], G.M.Ter-Akopian [e], A.V.Daniel [e], Y.Oganessian [e]

[a] Physics Department, Vanderbilt University, Nashville, TN 37235, USA

[b] Physics Department, Tsinghua University, Beijing 100084, People's Republic of China

[c] Lawrence Berkeley National Laboratory, Berkeley, CA 94720, USA

[d] Physics Department, University of Notre Dame, Notre Dame, IN 46556, USA

[e] Flerov Laboratory of Nuclear Reactions, JINR, Dubna, Russia

Pseudospin-doublet bands and Gallagher Moszkowski doublet bands in ^{100}Y

E. H. Wang, J. H. Hamilton, A. V. Ramayya, C. J. Zachary, A. Lemasson, A. Navin, M. Rejmund, S. Bhattacharyya, Q. B. Chen, S. Q. Zhang, J. M. Eldridge, J. K. Hwang, N. T. Brewer, Y. X. Luo, J. O. Rasmussen, S. J. Zhu, G. M. Ter-Akopian, Yu. Ts. Oganessian, M. Caamaño, E. Clément, O. Delaune, F. Farget, G. de France, and B. Jacquot
Phys. Rev. C **103**, 034301 – Published 2 March 2021

3. International Prestigious Awards and Honorary Degrees

Joe has received many prestigious awards and honorary degrees from many worldwide institutions, for example:

- Dr. Phil. Nat. Honoris Causa, Johann Wolfgang Goethe Universitat, Frankfurt, Germany, 1992.
- Dr. Phil. Nat. Honoris Causa — Univ. of Bucharest, Romania, 1999.
- Dr. Phil. Nat. Honoris Causa — St. Petersburg State University, Russia, 2001.
- Dr. Phil. Nat. Honoris Causa — Joint Institute for Nuclear Research, Russia, 2004.
- Dr. Phil. Nat. Honoris Causa — Shukla University, Raipur, India, 2005.
- Ilkovic Gold Medal for Career Achievement in Science, Slovak Academy of Sciences, 2001.
- Prize for International Scientific and Technological Cooperation with the People's Republic of China from PRC President, 2002.
- Flerov Prize for Major Achievements in Nuclear Physics Research, given by Russia, 2003.

4. The Joint Institute for Heavy Ion Research

The Joint Institute for Heavy Ion Research (JIHIR) has successfully engaged in more scientific international collaborations than any other group in the United States in the field of nuclear structure. In particular, JIHIR was responsible for a strong theoretical effort in low-energy nuclear physics that complemented the experimental programs in this area:

- Thousands of scientists and students (~100/year)
- Hundreds of meetings (~8/year)
- Summer schools

5. Summary

Let me extend my deep appreciation to you for the leadership you have shown within our community. As we look forward to exciting new nuclear physics — from studies of the structure of the nucleus to the study of nuclear collective motion and the myriad of applications of nuclear science — you have been and continue to be a real inspiration.

Thank You, Joe!

Joseph H. Hamilton: Global Outreach Educational Endeavors

A. Navin

*GANIL, CEA/DSM — CNRS/IN2P3, Bd Henri Becquerel,
BP 55027, F-14076 Caen Cedex 5, France*
navin@ganil.fr

In this contribution, we cover some of the amazing outreach and collaboration adventures of Joe. This is based on the presentation I gave at the Zoom meeting on Science and Humanity: The Extraordinary Life of Joseph H. Hamilton held on June 8th, 2022.

It's a great pleasure and honor to be part of the very special collection, but I am also very sad that I was not a student of Joe's and missed his lectures where he used scooters, rockets, and even a bed of nails! Here in this article, we try to give a flavor of some of the contributions to Joe's global outreach in research and education during the 64 years of his career. We are going around the world in 80 days and peep into his contribution to various countries and continents. I skip China as it will be described by Yixiao Luo. Without 80 days to go around the world, we make this voyage in the next few pages. Joe has not only had many pioneering and very exceptional contributions to US science but also has done a great many things outside the US. You have to forgive me as there are so many things to write home about so I had to pick only certain collaborations or countries to talk about. The above figure highlights some of the scientific aspects of his work. These include among others the successful search for the Super Heavy Elements (SHE) and study of cluster decay. The Hulk@Gammasphere represents what was used for Joe's work on fission fragment spectroscopy. The work on fission fragment spectroscopy will be discussed later. In each of the following figures, we try to give a feel about Joe's scientific collaborations and the accolades he received in various countries/continents.

1. Collaboration in Russia

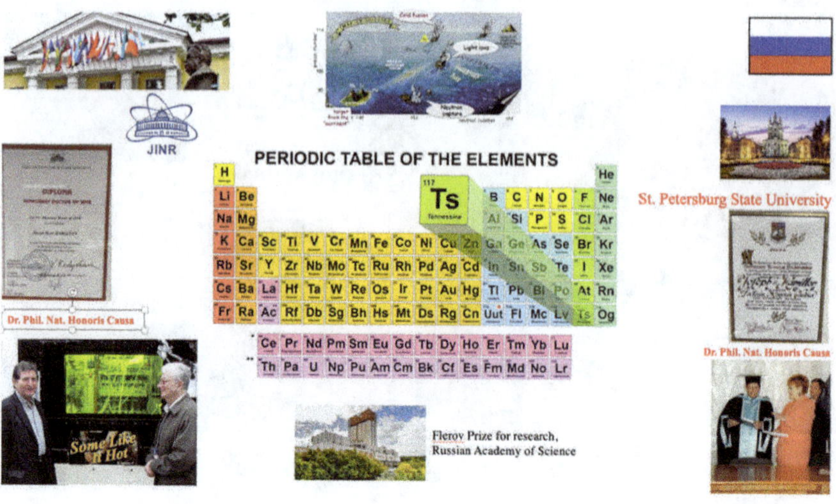

Joe had a fantastic, long-term relationship with Russia and especially with the Joint Institute of Nuclear Research in Dubna, the highlights of which are given in the figure above. The American–Russian combination worked together and new elements were discovered with key contributions, including the production of the target at ORNL where Joe played a major role. Another very fruitful collaboration, for the last 30 years, is related to studying the properties of neutron-rich nuclei, which has resulted in over 200 publications. On the bottom left of the above figure, Yuri "118" and Joe are near some "hot stuff" (target material) for their various research activities. The discovery of the new element, which Joe was given the honor of naming it for the state of Tennessee, "Tennessine" was also part of this program. He has received an honorary doctorate from Dubna and he is the only American to receive an honorary doctorate from St. Petersburg State University, Russia's most prestigious University. He also received, from the Russian Academy of Sciences, a prize for his research.

2. Collaborations with Romania and Slovak Republic

The next figure gives us an overview of his fruitful collaborations in both of these countries. He and Ramayya received honorary doctorates from Romania and Joe received a gold medal for his research from the Slovak

Academy of Science. The figure also shows the popularity of his (along with Fujia Yang) textbook based on the many years of teaching and research and is a crystallization of the intense passion and strong interest in the history of physics and the philosophy of science. He was also elected a foreign member of Academia Europaea a pan-European Academy of Humanities, Letters, Law, and Sciences.

3. Collaborations in Germany

Joe has had extensive collaborations with Goethe University of Frankfort and the major accelerator lab of Germany, GSI, near Frankfurt. First, he formed a long-term collaboration with Professor Walter Greiner in Frankfort. There they worked on a theoretical explanation of his discovery of shape coexistence in nuclei. He received an honorary doctorate from Goethe University and was awarded an Alexander Von Humboldt Prize by the President of Germany. Some of these are highlighted in the following figure. He spent a year at Goethe University, where he helped carry out the first Coulomb excitation of an actinide heavy element at GSI with a ^{248}Cm target he brought from Oak Ridge National Laboratory.

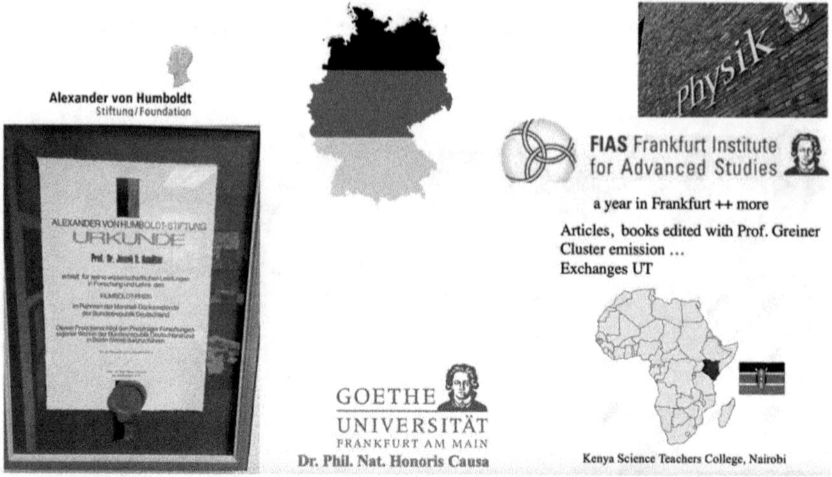

Professor Greiner also came to Vanderbilt every spring for 15 years to teach a theoretical physics course for graduate students and to do research with Joe. He would also bring two of his current PhD students with him. Also, he was selected to the Academy of Europe of Arts and Sciences. In addition, he twice spoke at conferences in then East Germany in Dresden.

Before I jump back to Europe, I want to just make a small transition to Africa. Joe has not left any continent untouched and taught at the Kenya Teachers College. He also communicated the beauty of his research at two conferences in South Africa and one in Namibia.

4. The French Connexion and More

Now, I come to France, where I am based. He spent a month as a guest Professor at the University in Strasbourg, He has given invited talks at conferences all over France. Arles, Caen, Chateau de Cadarache, Corsica, Divonne, Paris, San Malo, Seysins, and Strasbourg. He has similarly given invited talks all over Italy, from Sicily to Lake Como.

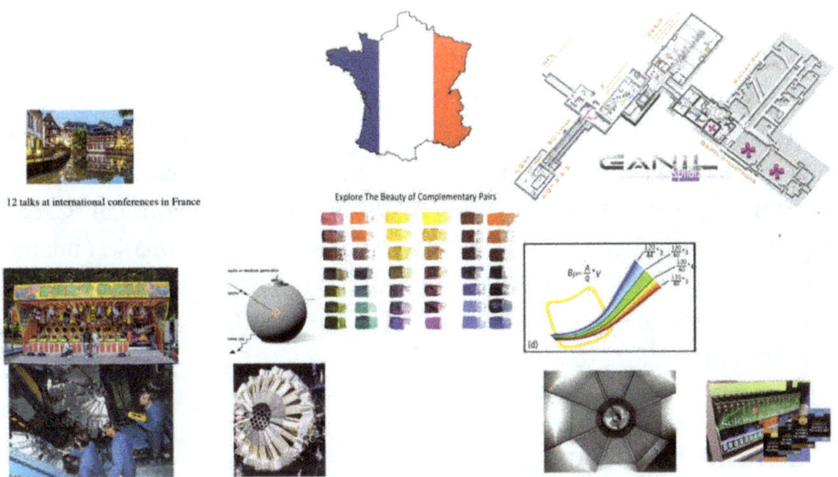

Now, I come back to what we mentioned earlier, namely probing the extremes of nuclear properties simultaneously at Spin *and* Isospin using the results of fission fragment spectroscopy. An overview of this work is given in the above figure. It was at a Sanibel conference where I met Joe. I gave a talk about our first results of the spectroscopy of fission fragments using EXOGAM γ-ray array combined with the VAMOS spectrometer.

Joe was immediately sharp enough to see how we could collaborate with him and his group to carry out new research neither of us could have done alone. Joe, as you know, was using Gammasphere to study neutron-rich nuclei up to very high spins using a ^{252}Cf spontaneous fission source. We were doing experiments bombarding uranium on beryllium to induce fission. With a spectrometer, we measured the A and Z of each of the fragments and their γ rays arising from low-lying excited states. These two sets of data contain extremely strong complementary data. I think this is a hallmark of Joe because he knows what you have and you realize the advantages of collaborating with other teams. We have published several long chains of new-level structures addressing a variety of questions in nuclear structure together. We have at least two more chains we are currently working on. A natural consequence of this collaboration, apart from the science impact, was that Joe made an important step towards FAIR, (**F**indability, **A**ccessibility, **I**nteroperability, and **R**euse of digital assets) for our field. The sharing of these "big" data sets in today's context has become very important in maximizing the science output for the money invested in science. In Europe and the United States, there are strong and important initiatives towards "FAIR".

Joe spent a year doing research in Sweden and a half year in the Netherlands and spoke at many conferences all over Italy and in Hungary, Romania, Bulgaria, Greece, Turkey, Yugoslavia, Spain, Portugal, Czech Republic, Slovak Republic, Poland, England, Scotland, Denmark, Finland, Netherlands, Belgium, and Russia in addition to France, Germany, India, and the far East and South America. He has also collaborated with scientists from most of these countries.

5. Indian Collaboration

I want to digress to my own country, so this one has a special place and I think it is very important for reasons below. Joe began his collaboration with India in 1964, when a young Indian physicist Suresh Pancholi came to Vanderbilt from Delhi to do research with Joe. He went back and became a key figure in molding Nuclear Physics at the Universities. Over the years, coming from India were many faculty and graduate students to do their PhD with Joe. The following figure gives a small glimpse of his connection with India. Together, with the Indian collaborators Joe published over 150 research papers.

The single most important contribution that India gave to Joe is AV Ramayya. He joined Joe as a research associate and continued on to join the faculty to become a full professor. Together they have formed a life-long collaboration and close friendship with over 700 research publications. If you want to do great science, you need at least one great collaborator. I think Joe and Ramayya have been such a pair. I also want to focus on the people who have worked with Joe. One young Indian J. J. Das came to train with Joe and ORNL and to contribute to bring their new recoil mass separator online as seen in the above figure. He went back to India and produced the first radioactive ion beam in India using the RMS at New Delhi. (The person on the left in the RMS photograph.) The other two in the photograph are with FRIB at Michigan State University.

6. Top Gun

In analogy with a recent movie of Tom Cruise, Joe is a Top Gun Maverick, who is not afraid of taking risks. The other thing the movie showed very

A dream is not what you see in your sleep
It is something which does not let you sleep

strongly was about collaborations and being a team. I think Joe has done both and what we see in the middle of the picture are the wings of Joe (fission fragment mass distributions).

Joe, it has been an honor to be associated with a legend like you. I have learnt where all one can make an impact and what is the real meaning of a contribution in the many important areas ranging from teaching to mentoring and to the growth of science. I tell the youngsters about your achievements contributions and accolades and also the fun stuff of you teaching during a baseball game and your scientific shows for your students. This is just to let them know that they have miles to go before they sleep, miles to go before they sleep. Thank you, Joe, and see you in Sanibel next year!

Joe Hamilton, International Cooperation, the Prisoner's Game, and Nuclear Structure

Ani Aprahamian[*,†], Kevin Lee[*], and Shelly R. Lesher[‡]

*Department of Physics & Astronomy, University of Notre Dame,
Notre Dame, IN 46556, USA

†A. Alikhanyan National Science Laboratory of Armenia,
Yerevan, 0036, Armenia

‡Department of Physics, University of Wisconsin-La Crosse,
La Crosse, WI 54601, USA

We bring forward glimpses of the man via interactions with and impressions of Joe Hamilton over four decades starting with graduate school, the road to academia, and further travel to foreign lands particularly to the Russian Federation. Joseph H. Hamilton has been an outstanding example of the ideal scientist without borders, motivated by science and friendships resulting in the mutual benefit to all sides. The collaboration of The Joint Institute for Nuclear Research (JINR) with Oak Ridge National Laboratory (ORNL) enabled by Joe led to the discovery of new elements in the last row of the periodic table and the naming of an element especially for the state of Tennessee (Tennessine: Ts). In this tribute to his "Extraordinary Life", we share progress in the area of nuclear structure that was near and dear to the heart of Joe Hamilton, fission, as well as the progress in the creation of new elements.

1. Introduction

Joseph H. Hamilton and I share a birthday in the middle of August some twenty-six years apart. What a delightful surprise to get a call from Joe Hamilton a few years ago on our birthday with a request and wish to visit the Sanford Underground Research Facility (SURF) in South Dakota. The underground physics research at SURF includes measuring, at very low energies, reactions of relevance to stellar evolution, the search for dark matter, and the precision measurement of neutrino mixing via CP violation to determine the neutrino mass hierarchy. These are all topics quite far from the research foci of Joe Hamilton but his curiosity is boundless for new discoveries and has been an active participant in teaching and research into his 90s.

Joseph Hamilton also lived an exemplary life in collaborations and friendships. Even though he was probably never familiar with game theory, the "prisoner game", or the vast numbers of theoretical papers on international cooperation and collaboration in politics,[1] his life was living proof of those theories. The rules of the games are simple and they include evaluations of the benefits to the individual participants if they fully, partially or not-at all cooperate. The results in simple words show maximum benefit if all players collaborate and less so with fractional or no cooperation.

Friendship and collaboration resulted in the completion of the last row of the periodic table (shown in Fig. 1) and involved a close collaboration of the USA (ORNL, LLNL), Russian Federation (JINR, FLNR), Japan (RIKEN), and Germany (GSI). The efforts resulted joint efforts in search of superheavy nuclei was a transatlantic or trans-global effort.[2] The recent advances at JINR with the development of the Dubna Gas-Filled Recoil Separator 2 at the Superheavy Element Factory[3] have not only made advances in measuring the decay paths of all the last row of elements but also have some indication that these elements are perhaps not the last elements of the periodic table.[4-6]

The main goals of Joseph Hamilton's research was nuclear structure. Structure of neutron-rich nuclei from spontaneous fission, high-spin excitations of nuclei from heavy ion reactions, as well as the search for the island of stability. In structure, Joe Hamilton was current and engaged

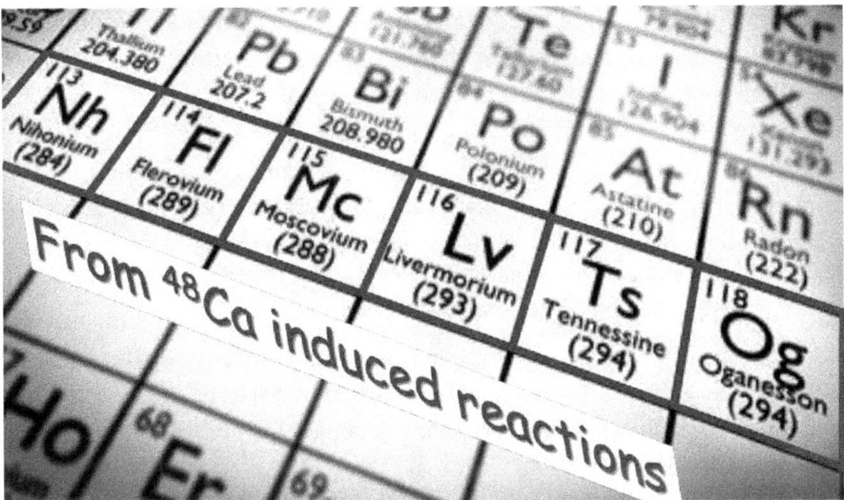

Fig. 1.   The last elements completing the periodic table of elements including Nihonium ($Z = 113$), Flerovium ($Z = 114$), Moscovium ($Z = 115$), Livermorium ($Z = 116$), Tennessine ($Z = 117$), and Oganesson ($Z = 118$) produced in hot fusion reactions using ^{48}Ca. FLNR pioneered reactions between 48Ca beams and enriched actinide targets that led to the new element discoveries.

with theoretical developments in nuclear theory. The 1975 Nobel Prize in Physics to Bohr, Mottelson, and Rainwater for the discovery of the connection between nucleon motion and the emergent collective behavior inspired research in search of those predicted low-lying oscillations of the nucleus. The studies from spontaneous fission, the heavy ion reactions, and the superheavy element productions are all connected through the love of Joe Hamilton for understanding the true nature of nuclear matter.

Although the scientific contributions of Joe Hamilton are significant, it is very worthwhile to mention the generosity and kindness of the man behind it all. We were overwhelmed with the kindness of Joseph Hamilton when I had been struggling to publish some measurements of lifetimes for the ^{178}Hf nucleus. We were turned down from publication in *Physical Review Letters* after several attempts with strong guidance of the reviewer to remeasure certain aspects of the work that we were trying to present. We repeated the experiments three times and received the same results.

Each time, we attempted to convince the reviewer and failed to do so. It was an act of great kindness on the part of Joe Hamilton that saved us. Hamilton phoned me and revealed that he was the reason that our work was being blocked from publication. He explained his reasons. He had studied the same nucleus and did not find the states whose lifetimes we were reporting. He admitted that it was the dissertation of one of his graduate students. We were persistent, we tracked down the graduate student who was now a staff scientist and re-examined his data. We jointly discovered that just in the region in question, there had been a mistake in the calibration of the spectrum. We could only then persuade Joe and published the work.[7] We have never forgotten this act of kindness to explain why a manuscript was not being accepted for publication. Dare we say that such generosity is so extremely rare.

In the following, we briefly discuss Joe's research in nuclear structure of neutron rich nuclei produced by spontaneous fission of ^{252}Cf and developments in nuclear structure that tie all of his work together from super-heavies, to fission, and to vibrational excitations.

2. Fission

Fission of nuclei has been studied in nuclear science from the birth of the field. However, fission and fission distributions of neutron rich matter has emerged as a leading question in nuclear science.[8] The arrival on earth of the gravitational waves (GW 170817) resulting from the merger of two-neutron stars 132 million years ago and the 70 electromagnetic satellites pointing to the source showed the creation of the elements. Figure 2 shows the evolution of light in time starting with the lighter elements in blue and eventually going into the Infra Red by the synthesis of the rare-earth elements. Joe Hamilton's work was in large part based on the spectroscopy of nuclei produced in spontaneous fission of ^{252}Cf. This fission process gave access to the production of neutron rich nuclei and allowed the exploration of the similarities and differences in structure of neutron rich nuclei with those closer to stability with γ-ray spectroscopy.

While a two-neutron star merger was identified as a source for the synthesis of the heavy elements with rapid neutron capture, detailed questions remain about fission and fission recycling. Questions on the

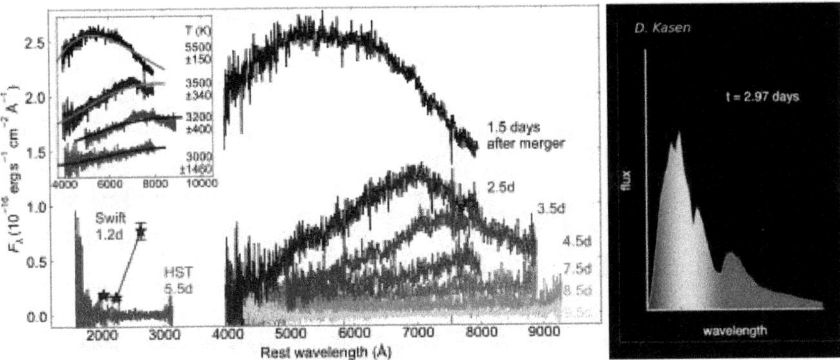

Fig. 2.   The wavelengths of light seen from the source of the gravitational waves resulting from the merger of two-neutron stars as a function of time on the left. On the right, the compiled light from the synthesis of the elements into the deep red (lanthanides).[9]

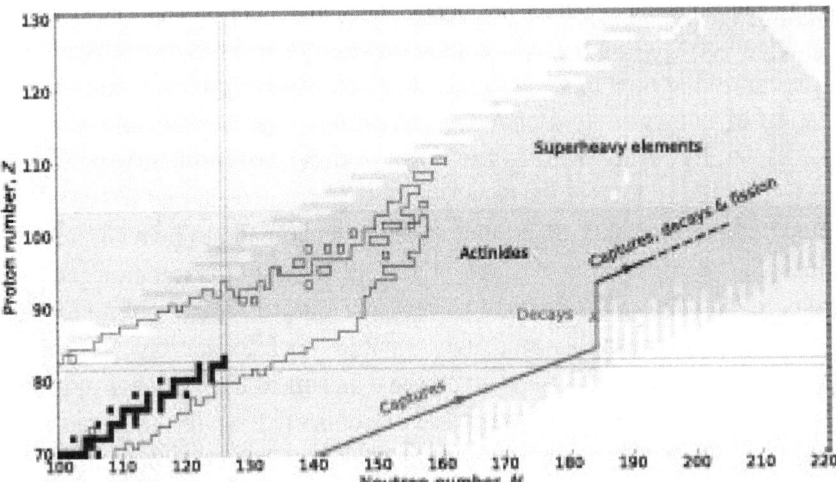

Fig. 3.   The chart of nuclides in the actinide region with indication of the r-process path.[10] This path is far from the expected island of stability and the superheavy element region. However, data are limited regarding cluster decay in the region and the possibility to populate the superheavy nuclei in the r-process.

production of superheavy elements in nature by the r-process is also likely.[10–13] Figure 3 connects the r-process path with the island of superheavy elements. These new developments build on studies of Joe with the production of the heavy elements, fission, and nuclear structure.

3. Nuclear Structure

The 1975 Nobel Prize in Physics was awarded to Bohr, Mottelson, and Rainwater describing nuclei geometrically as a shape and the oscillations of the nucleus around that shape. The lowest lying shape effecting oscillations or vibrations in this description would be quadrupole ($\lambda = 2$) in nature, resulting in two types of vibrations in deformed nuclei: β with oscillations along the symmetry axis ($K^{\pi} = 0^{+}$) and γ breaking axial symmetry with a projection of $K^{\pi} = 2^{+}$ on the symmetry axis. The γ vibration seems to be well characterized as the first $K^{\pi} = 2_{1}^{+}$ (or 2_{γ}^{+}) band and exhibits a systematic behavior across the region of deformed nuclei with typical $B(E2 : 2_{\gamma}^{+} \rightarrow 0_{g.s.}^{+})$ values of a few Weisskopf units (W.u.).[14] Today, over forty-five years later, it is the existence and characterization of the low-lying β vibration which remains an open question in nuclear structure.[7,14–47]

Joe Hamilton has been engaged in experiments and discoveries of single and double-phonon vibrational excitations of both quadrupole and octupole oscillations in several nuclei.[48–52] His work has led to the identification of numerous two-phonon γ-vibrational states in several neutron rich nuclei, frequently studied by γ-ray spectroscopy following spontaneous fission of ^{252}Cf. Over the decades, it has become clear that 0^{+} states are populated more weakly than other states in fission and in heavy ion reactions. The identification of single and double-phonon excitations of the β-type have been more difficult to measure due to the lack of data. The observation of strongly collective β vibrational excitations in nuclei appear to be somewhat dependent on the experimental techniques or tools. The existence of strong and consistently collective γ-vibrational excitations without its complement β-vibrational states would present some unique challenges for quantum mechanics in nuclei making them unique and different from atoms or molecules. In well-deformed regions of nuclei, excitations built on a deformed ground state have traditionally been described in terms of quadrupole excitations leading to the classifications of the first excited 0^{+} bands as single-phonon β-vibrational excitations. However, discussions in recent years have focused on a debate about the absence, or lack of, a ($K^{\pi} = 0^{+}$) β vibration with a multitude of possible interpretations including the possibility of phase changes at the onset of deformation (for example, at $N = 90$ and $Z = 64$) and the application of new symmetries to describe these nuclei,[17,53–58] or the change in the expectation of a β vibration. The discussions on the existence or absence of the $K^{\pi} = 0^+$ β vibrational excitations in nuclei have spanned a wide spectrum of possibilities from shape co-existence, where a competing shape is not the lowest favored shape but occurs low in the excitation spectrum of a given nucleus,[15] to a redefinition of what can, in fact, be interpreted as a vibration.[59] In the IBM,[60–62] the first excited 0^+ and 2^+ bands are members of the same representation and in a pure SU(3) limit, would not decay to the ground state (g.s.). Most deformed nuclei however are not pure SU(3) and the breaking of that symmetry gives rise to interband transitions from both of the low-lying $K^{\pi} = 2^+$ and $K^{\pi} = 0^+$ (and) bands of significant strengths. Another development describes nuclei at the point of phase change from spherical to deformed in terms of β and γ shape parameters, or the SU(3) symmetry[30,31,33] or the pseudo-SU(3).[47] There is also the possibility of Partial Dynamical Symmetries where the SU(3) symmetry is obeyed by some of the states and broken in others.[63] A systematic theoretical study of even-even deformed nuclei in the Hartree-Fock-Bogoliubov approach extended by the generator coordinator method and mapped into a five-dimensional collective Hamiltonian for even-even nuclei from $Z = 10 - 110$, provides guidelines to distinguishing between coexistence and β vibrational oscillations. These studies of the nature of $K^{\pi} = 0^+$ bands in deformed nuclei predict widely varying levels of collectivity depopulating the first excited 0^+ states.[14]

The activity and fervor regarding the nature of low-lying 0^+ states in nuclei has re-invigorated the debate about the nature of low-lying 0^+ excitations in the spectra of nuclei, particularly in light of the results of high precision (p,t) measurements from the Q3D at the University of Munich.[21,22] Figure 4 summarizes the results of recent work and show that there are tens of 0^+ states in the spectra of deformed rare earth nuclei. Many of the 0^+ states are even below the pairing gaps.

The underlying science question has remained unanswered. In nuclei, where there is explicit deformation in the ground state, "are the low-lying 0^+ states collective vibrations built on the ground state or are they minima of a coexisting shape?" Delaroche *et al.*[64] have shown that for a significant percentage of $K = 0^+$ excitations built on the deformed g.s. should in fact be collective vibrations.

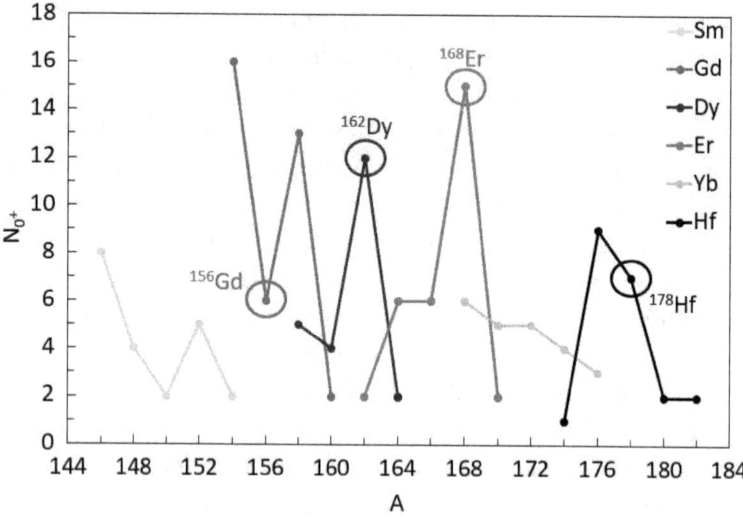

Fig. 4. (Color online) The total number of confirmed and tentative 0^+ states observed in several deformed rare-earth nuclei (Sm, Gd, Dy, Er, Yb, and Hf).

We have followed the remarkable developments in the identification of 0^+ states and determined to understand their character, we have measured lifetimes of the observed 0^+ states. From the results of our work, we present some examples of double-phonon $\gamma\gamma$, $\beta\beta$, and a $\beta\gamma$ vibrations. What remains for a complete understanding of the nature of the observed 0^+ states, following the transfer reactions and lifetime measurements, are the $E0$ measurements.

The results of the lifetime measurements for several nuclei in the rare earth region have been studied and interesting characteristics emerge for the 0^+ bands. In the ^{156}Gd, there are two low-lying $K^\pi = 0^+$ bands. The lower 0^+ band at 1049.5 keV carries the transitions with the greater collective strength to the ground state while there is a higher lying $K^\pi = 0^+$ band at 1715.1 keV which is strongly connected to the γ band or the first excited $K^\pi = 2^+$ band. The transition probabilities imply that it is a two-phonon $\gamma\gamma$ vibration. In the ^{162}Dy nucleus, we find the fragmented strength of two-phonon $\gamma\gamma$ vibration and for the first time, a two-phonon $\beta\gamma$ vibration. In the ^{168}Er nucleus, there are several 0^+ states the one with the greater transition probabilities connecting to the ground state is the third excited $K^\pi = 0^+$

Table 1. Energy and B($E2$) ratios identified in lifetime studies in comparison to theoretical values. There are two values if fragmentation is observed and measured.

Ratios	^{156}Gd	^{162}Dy	^{168}Er	Theory
$E(4^+_{\gamma\gamma})/E(2^+_\gamma)$	1.31	1.73; 2.5	2.5	2.0
$E(0^+_{\gamma\gamma})/E(2^+_\gamma)$	1.49	1.6		2.0
$E(2^+_{\beta\gamma})/E(2^+_\gamma)$		2.5		≈ 2.0
$B(E2{:}4^+_{\gamma\gamma} \to 2^+_\gamma)/B(E2{:}2^+_\gamma \to 0^+_{g.s})$	0.39	14 ± 9; 1.8		2.78
$B(E2{:}0^+_{\gamma\gamma} \to 2^+_\gamma)/B(E2{:}2^+_\gamma \to 0^+_{g.s})$	1.96			5.0
$B(E2{:}2^+_{\beta\gamma} \to 2^+_\gamma)/B(E2{:}2^+_\gamma \to 0^+_{g.s})$		1.1		1.43

band. We summarize some of the information on single and double-phonon excitations in Table 1 in comparison with expected theoretical values.

4. Conclusions

The path of Joe Hamilton sketched through nuclear structure, fission, and the production of heavy elements leads us forward. We intend to get to the answers about the viability of low-lying collective oscillations in a comprehensive way, to develop signatures that will allow us to distinguish between coexistence and collective vibrations. Even further, perhaps with the determinations of the decay chains for the heavy elements, we will answer the question about the production of the superheavies in nature.

References

1. S. Barrett, A Theory of Full International Cooperation, *J. Theor. Politics*, **11**(4), 519–541 (1999). doi: 10.1177/0951692899011004004.

2. J. Hamilton, S. Hofmann, and Y. Oganessian, Search for Superheavy Nuclei, *Annu. Rev. Nucl. Part. Sci.*, **63**(1), 383–405 (2013). doi: 10.1146/annurev-nucl-102912-144535.

3. D. Solovyev and N. Kovrizhnykh, Simulations of Recoil Trajectories in Dubna Gas-Filled Recoil Separator 2 by Geant4 Toolkit, *JINST*, **17**(07), P07033 (2022). https://dx.doi.org/10.1088/1748-0221/17/07/P07033.

4. Y. T. Oganessian, V. K. Utyonkov, N. D. Kovrizhnykh, F. S. Abdullin, S. N. Dmitriev, D. Ibadullayev, M. G. Itkis, D. A. Kuznetsov, O. V. Petrushkin, A. V. Podshibiakin, A. N. Polyakov, A. G. Popeko, R. N. Sagaidak, L. Schlattauer, I. V. Shirokovski, V. D. Shubin, M. V. Shumeiko, D. I. Solovyev, Y. S. Tsyganov, A. A. Voinov, V. G. Subbotin, A. Y. Bodrov, A. V. Sabel'nikov, A. V. Khalkin, V. B. Zlokazov, K. P. Rykaczewski, T. T. King, J. B. Roberto, N. T. Brewer, R. K. Grzywacz, Z. G. Gan, Z. Y. Zhang, M. H. Huang, and H. B. Yang, First experiment at the super heavy element factory: High cross section of ^{288}Mc in the ^{243}Am + ^{48}Ca reaction and identification of the new isotope ^{264}Lr, *Phys. Rev. C.*, **106**, L031301 (2022). https://link.aps.org/doi/10.1103/PhysRevC.106.L031301.

5. Y. T. Oganessian, V. K. Utyonkov, D. Ibadullayev, F. S. Abdullin, S. N. Dmitriev, M. G. Itkis, A. V. Karpov, N. D. Kovrizhnykh, D. A. Kuznetsov, O. V. Petrushkin, A. V. Podshibiakin, A. N. Polyakov, A. G. Popeko, R. N. Sagaidak, L. Schlattauer, V. D. Shubin, M. V. Shumeiko, D. I. Solovyev, Y. S. Tsyganov, A. A. Voinov, V. G. Subbotin, A. Y. Bodrov, A. V. Sabel'nikov, A. Lindner, K. P. Rykaczewski, T. T. King, J. B. Roberto, N. T. Brewer, R. K. Grzywacz, Z. G. Gan, Z. Y. Zhang, M. H. Huang, and H. B. Yang, Investigation of ^{48}Ca-induced reactions with ^{242}Pu and ^{238}U targets at the jinr superheavy element factory, *Phys. Rev. C.*, **106**, 024612 (2022). https://link.aps.org/doi/10.1103/ PhysRevC.106.024612.

6. Y. T. Oganessian, V. K. Utyonkov, N. D. Kovrizhnykh, F. S. Abdullin, S. N. Dmitriev, A. A. Dzhioev, D. Ibadullayev, M. G. Itkis, A. V. Karpov, D. A. Kuznetsov, O. V. Petrushkin, A. V. Podshibiakin, A. N. Polyakov, A. G. Popeko, I. S. Rogov, R. N. Sagaidak, L. Schlattauer, V. D. Shubin, M. V. Shumeiko, D. I. Solovyev, Y. S. Tsyganov, A. A. Voinov, V. G. Subbotin, A. Y. Bodrov, A. V. Sabel'nikov, A. V. Khalkin, K. P. Rykaczewski, T. T. King, J. B. Roberto, N. T. Brewer, R. K. Grzywacz, Z. G. Gan, Z. Y. Zhang, M. H. Huang, and H. B. Yang, New isotope ^{286}Mc produced in the ^{243}Am + ^{48}Ca reaction, *Phys. Rev. C.*, **106**, 064306 (2022). https://link.aps.org/doi/10.1103/ PhysRevC.106.064306.

7. A. Aprahamian, R. C. de Haan, H. G. Börner, H. Lehmann, C. Doll, M. Jentschel, A. M. Bruce, and R. Piepenbring, Lifetime measurements of excited $K^{\pi} = 0^{+}$ bands in ^{178}Hf, *Phys. Rev. C.*, **65**, 031301(R) (2002).

8. A. Aprahamian, Open challenges to nuclear physics resulting from the neutron star merger, *Nucl. Phys. News*, **31**(3), 11–16 (2021). doi: 10.1080/10619127.2021.1915019.

9. M. Nicholl, E. Berger, D. Kasen, B. D. Metzger, J. Elias, C. Briceño, K. D. Alexander, P. K. Blanchard, R. Chornock, P. S. Cowperthwaite, T. Eftekhari,

W. Fong, R. Margutti, V. A. Villar, P. K. G. Williams, W. Brown, J. Annis, A. Bahramian, D. Brout, D. A. Brown, H.-Y. Chen, J. C. Clemens, E. Dennihy, B. Dunlap, D. E. Holz, E. Marchesini, F. Massaro, N. Moskowitz, I. Pelisoli, A. Rest, F. Ricci, M. Sako, M. Soares-Santos, and J. Strader, The electromagnetic counterpart of the binary neutron star merger ligo/virgo gw170817. iii. optical and uv spectra of a blue kilonova from fast polar ejecta, *ApJL*, **848**(2), L18 (2017). https://dx.doi.org/10.3847/2041-8213/aa9029.

10. E. M. Holmbeck, T. M. Sprouse and M. R. Mumpower, Nucleosynthesis and observation of the heaviest elements, *Eur. Phys. J. A.*, **59**(2), 28 (2023). https://doi.org/10.1140/epja/s10050-023-00927-7.

11. I. Petermann, K. Langanke, G. Martínez-Pinedo, I. V. Panov, P.-G. Reinhard and F.-K. Thielemann, Have superheavy elements been produced in nature?, *Eur. Phys. J. A.*, **48**(9), 122 (2012). https://doi.org/10.1140/epja/i2012-12122-6.

12. Z. Matheson, S. A. Giuliani, W. Nazarewicz, J. Sadhukhan, and N. Schunck, Cluster radioactivity of $^{294}_{118}$Og$_{176}$, *Phys. Rev. C.*, **99**, 041304 (2019). https://link.aps.org/doi/10.1103/PhysRevC.99.041304.

13. N. Vassh, R. Vogt, R. Surman, J. Randrup, T. M. Sprouse, M. R. Mumpower, P. Jaffke, D. Shaw, E. M. Holmbeck, Y. Zhu, and G. C. McLaughlin, Using excitation-energy dependent fission yields to identify key fissioning nuclei in r-process nucleosynthesis, *JPhysG*, **46**(6), 065202 (2019). https://dx.doi.org/10.1088/1361-6471/ab0bea.

14. A. Aprahamian, S. R. Lesher, C. Casarella, H. G. Börner, and M. Jentschel, Lifetime measurements in ^{162}Dy, *Phys. Rev. C.*, **95**, 024329 (2017).

15. K. Heyde and J. L. Wood, Shape coexistence in atomic nuclei, *Rev. Mod. Phys.*, **83**, 1467 (2011).

16. P. E. Garrett, W. D. Kulp, J. L. Wood, D. Bandyopadhyay, S. Choudry, D. Dashdorj, S. R. Lesher, M. T. McEllistrem, M. Mynk, J. N. Orce, and S. W. Yates, New features of shape coexistence in ^{152}sm, *Phys. Rev. Lett.*, **103**, 062501 (2009).

17. R. F. Casten and P. von Brentano, New interpretation of the lowest k = 0 collective excitation of deformed nuclei as a phonon excitation of the γ band, *Phys. Rev. C.*, **50**, R1280 (1994).

18. A. Aprahamian, R. C. de Haan, S. R. Lesher, J. Döring, A. M. Bruce, H. G. Börner, M. Jentschel, and H. Lehmann, Collective $K^{\pi} = 0^{+}$ vibrational excitations in ^{178}Hf, *J. Phys.*, **G25**, 685 (1999). doi: 10.1088/0954-3899/25/4/020.

19. X. Wu, A. Aprahamian, J. Castro-Ceron, and C. Baktash, Identical bands and multi-phonon vibrations, *Phys. Lett. B.*, **316**, 235 (1993). doi: 10.1016/0370-2693(93)90319-D.

20. X. Wu, A. Aprahamian, S. M. Fischer, W. Reviol, G. Liu, and J. X. Saladin, Multiphonon vibrational states in deformed nuclei, *Phys. Rev. C.*, **49**, 1837 (1994).

21. S. R. Lesher, A. Aprahamian, L. Trache, A. Oros-Peusquens, S. Deyliz, A. Gollwitzer, R. Hertenberger, B. D. Valnion, and G. Graw, New $K^\pi = 0^+$ bands in ^{158}Gd, *Phys. Rev. C.*, **66**, 051305(R) (2002).

22. D. A. Meyer, V. Wood, R. F. Casten, C. R. Fitzpatrick, G. Graw, D. Bucurescu, J. Jolie, P. von Brentano, R. Hertenberger, H.-F. Wirth, N. Braun, T. Faestermann, S. Heinze, J. L. Jerke, R. Krücken, M. Mahgoub, O. Möller, D. Mücher, and C. Scholl, Extensive investigation of 0^+ states in rare earth region nuclei, *Phys. Rev. C.*, **74**, 044309 (2006).

23. D. Bucurescu, G. Graw, R. Hertenberger, H. Wirth, N. L. Iudice, A. V. Sushkov, N. Y. Shirikova, Y. Sun, T. Faestermann, R. Krücken, M. Mahgoub, J. Jolie, P. von Brentano, N. Braun, S. Heinze, O. Möller, D. Mücher, C. Scholl, R. F. Casten, and D. A. Meyer, High-resolution study of 0^+ and 2^+ excitations in ^{168}er with the (p,t) reaction, *Phys. Rev. C.*, **73**, 064309 (2006).

24. S. R. Lesher, J. N. Orce, Z. Ammar, C. D. Hannant, M. Merrick, N. Warr, T. B. Brown, N. Boukharouba, C. Fransen, M. Scheck, M. T. McEllistrem, and S. W. Yates, Study of 0^+ excitations in ^{158}Gd with the $(n,n'\gamma)$ reaction, *Phys. Rev. C.*, **76**, 034318 (2007).

25. L. Bettermann, S. Heinze, J. Jolie, D. Mücher, O. Möller, C. Scholl, R. F. Casten, D. A. Meyer, G. Graw, R. Hertenberger, H.-F. Wirth, and D. Bucurescu, High-resolution study of 0^+ states in ^{170}yb, *Phys. Rev. C.*, **80**, 044333 (2009). doi: 10.1103/PhysRevC.80.044333.

26. C. Bernards, R. F. Casten, V. Werner, P. von Brentano, D. Bucurescu, G. Graw, S. Heinze, R. Hertenberger, J. Jolie, S. Lalkovski, D. A. Meyer, D. Mücher, P. Pejovic, C. Scholl, and H.-F. Wirth, High-resolution study of excited 0^+ states in ^{200}hg and ^{202}hg, *Phys. Rev. C.*, **87**, 064321 (2013). doi:10.1103/PhysRevC.87.064321.

27. C. Bernards, R. F. Casten, V. Werner, P. von Brentano, D. Bucurescu, G. Graw, S. Heinze, R. Hertenberger, J. Jolie, S. Lalkovski, D. A. Meyer, D. Mücher, P. Pejovic, C. Scholl, and H.-F. Wirth, Investigation of 0^+ states in ^{198}hg after two-neutron pickup, *Phys. Rev. C.*, **87**, 024318 (2013). doi: 10.1103/PhysRevC.87.024318.

28. N. V. Zamfir, J.-y. Zhang, and R. F. Casten, Interpreting recent measurements of 0^+ states in ^{158}gd, *Phys. Rev. C.*, **66**, 057303 (2002). doi: 10.1103/PhysRevC.66.057303.

29. Y. Sun, A. Aprahamian, J. Zhang, and C.-T. Lee, Nature of excited 0^+ states in ^{158}gd described by the projected shell model, *Phys. Rev. C.*, **68**, 061301(R) (2003). doi: 10.1103/PhysRevC.68.061301.

30. N. Pietralla and O. M. Gorbachenko, Evolution of the "β excitation" in axially symmetric transitional nuclei, *Phys. Rev. C.*, **70**, 011304(R) (2004).

31. K. Dusling, N. Pietralla, G. Rainovski, T. Ahn, B. Bochev, A. Costin, T. Koike, T. C. Li, A. Linnemann, S. Pontillo, and C. Vaman, Mediumspin γ-ray spectroscopy of the transitional nucleus ^{160}er, *Phys. Rev. C.*, **73**, 014317 (2006).

32. R. Fossion, C. E. Alonso, J. M. Arias, L. Fortunato, and A. Vitturi, Shapephase transitions and two-particle transfer intensities, *Phys. Rev. C.*, **76**, 014316 (2007). doi: 10.1103/PhysRevC.76.014316.

33. D. Bonatsos, E. A. McCutchan, R. F. Casten, R. J. Casperson, V. Werner, and E. Williams, Regularities and symmetries of subsets of collective 0^+ states, *Phys. Rev. C.*, **80**, 034311 (2009).

34. D. Bonatsos, I. E. Assimakis, N. Minkov, A. Martinou, S. Sarantopoulou, R. B. Cakirli, R. F. Casten, and K. Blaum, Analytic prediction for nuclear shapes, prolate dominance, and the prolate-oblate shape transition in the proxy-su(3) model, *Phys. Rev. C.*, **95**, 064326 (2017). doi: 10.1103/PhysRevC.95.064326.

35. D. Bonatsos, I. E. Assimakis, N. Minkov, A. Martinou, R. B. Cakirli, R. F. Casten, and K. Blaum, Proxy-su(3) symmetry in heavy deformed nuclei, *Phys. Rev. C.*, **95**, 064325 (2017). doi: 10.1103/PhysRevC.95.064325.

36. R. M. Clark, R. F. Casten, L. Bettermann, and R. Winkler, Unified framework for understanding pair transfer between collective states in atomic nuclei, *Phys. Rev. C.*, **80**, 011303(R) (2009). doi: 10.1103/PhysRevC.80.011303.

37. N. Lo Iudice, V. Yu. Ponomarev, Ch. Stoyanov, A. V. Sushkov, and V. V. Voronov, Low-energy nuclear spectroscopy in a microscopic multiphonon approach, *J. Phys. G: Nucl. Part. Phys.*, **39**, 043101 (2012).

38. F.-Q. Chen, Y. Sun, and P. Ring, Quantum fluctuations in the collective 0^+ states of deformed nuclei, *Phys. Rev. C.*, **88**, 014315 (2013).

39. P. E. Garrett, M. Kadi, C. A. McGrath, V. Sorokin, M. Li, M. Yeh, and S. W. Yates, The nature of 0^+ excitations in ^{166}er, *Phys. Lett. B.*, **400**, 250 (1997).

40. R. C. de Haan, A. Aprahamian, H. G. Börner, C. Doll, M. Jentschel, A. M. Bruce, and S. R. Lesher, Lifetime measurements in ^{178}Hf, *J. Res. Natl. Inst. Stand. Technol.*, **105**, 125 (2000).

41. A. Aprahamian, The nature of low-lying $K^\pi = 0^+$ bands in nuclei, *Phys. Atom. Nucl.*, **67**, 1750 (2004). doi: 10.1134/1.1806918.

42. J. F. Sharpey-Schafer, T. E. Madiba, S. P. Bvumbi, E. A. Lawrie, J. J. Lawrie, A. Minkova, S. M. Mullins, P. Papka, D. G. Roux, and Timár, Blocking of coupling to the 0^+_2 excitation in ^{154}gd by the [505]11/2- neutron in ^{155}gd, *Eur. Phys. J. A.*, **47**, 6 (2011).

43. J. F. Sharpey-Schafer, The structure of excied 0^+ states in nuclei and the effect of the γ degree of freedom, *AIP Conf. Proc.*, **1377**, 205 (2011).

44. J. F. Sharpey-Schafer, S. M. Mullins, R. A. Bark, J. Kau, F. Komati, E. A. Lawrie, J. J. Lawrie, T. E. Madiba, P. Maine, A. Minkova, S. H. T. Murray, N. J. Ncapayi, and P. A. Vymers, Congruent band structures in ^{154}gd: Configuration-dependent pairing, a double vacuum and lack of β-vibrations, *Eur. Phys. J. A.*, **47**, 5 (2011).

45. T. Papenbrock and H. A. Weidenmüller, Effective field theory for deformed atomic nuclei, *Phys. Scr.*, **91**, 053004 (2016). doi: 10.1088/0031-8949/91/5/053004.

46. E. A. Coello Pérez and T. Papenbrock, Effective theory for nonrigid rotor in an electromagnetic field: Toward accurte and precise calculations of $e2$ transitions in deformed nuclei, *Phys. Rev. C.*, **92**, 014323 (2015). doi: 10.1103/PhysRevC.92.014323.

47. G. Popa, J. G. Hirsch, and J. P. Draayer, Shell model description of normal parity bands in even-even heavy deformed nuclei, *Phys. Rev. C.*, **62**, 064313 (2000).

48. C. Yong-Jing, Octupole deformed band and quasi-gamma band in ce-146 nucleus, *High Energy Physics and Nuclear Physics.*, **30**, 740 (2006).

49. S. Zhu, Y. Luo, J. Hamilton, J. Rasmussen, A. Ramayya, J. Hwang, H. Ding, X. Che, Z. Jiang, P. Gore, E. Jones, K. Li, I. Lee, W. Ma, G. Ter-Akopian, A. Daniel, S. Frauendorf, V. Dimitrov, J. Zhang, A. Gelberg, I. Stefancescu, and J. Cole, Triaxiality, chiral bands and gamma vibrations in a=99 – 114 nuclei, *Prog. Part. Nucl. Phy.*, **59**(1), 329–336 (2007). https://www.science-direct.com/science/article/pii/S0146641006001025.

50. J.-G. Wang, S.-J. Zhu, J. Hamilton, A. Ramayya, J. Hwang, S. Liu, K. Li, Y. Luo, J. Rasmussen, I. Lee, H.-B. Ding, Q. Xu, L. Gu, E.-Y. Yeoh, Z.-G. Xiao, and W. Ma, Identification of one-phonon and two-phonon γ-vibrational bands in odd-z 103nb nucleus, *Phys. Lett. B.*, **675**(5), 420–425 (2009). https://www.sciencedirect.com/science/article/pii/S0370269309004845.

51. D. Huai-Bo, Z. Sheng-Jiang, J. H. Hamilton, A. V. Ramayya, J. K. Hwang, Y. X. Luo, J. O. Rasmussen, I. Y. Lee, C. Xing-Lai, C. Yong-Jing, and L. Ming-Liang, Search for double γ-vibrational bands in neutron-rich ^{105}mo nucleus, *Chinese Phys. Lett.*, **23**(12), 3222 (2006). https://dx.doi.org/10.1088/0256-307X/23/12/028.

52. J. Marcellino, E. H. Wang, C. J. Zachary, J. H. Hamilton, A. V. Ramayya, G. H. Bhat, J. A. Sheikh, A. C. Dai, W. Y. Liang, F. R. Xu, J. K. Hwang, N. T. Brewer, Y. X. Luo, J. O. Rasmussen, S. J. Zhu, G. M. Ter-Akopian, and

Y. T. Oganessian, One- and two-phonon γ-vibrational bands in neutron-rich ^{107}Mo, *Phys. Rev. C.*, **96**, 034319 (2017). https://link.aps.org/doi/10.1103/PhysRevC.96.034319.

53. A. Arima and F. Iachello, Interacting boson model of collective states. i. the vibrational limit, *Ann. Phys. (NY).*, **99**, 253 (1976).

54. R. F. Casten, P. von Brentano, and N. V. Zamfir, Robust predictions of the interacting boson approximation model, *Phys. Rev. C.*, **49**, 1940 (1994).

55. R. F. Casten and P. von Brentano, Reply to "comment on 'new interpretation of the lowest $k = 0$ collective excitation of deformed nuclei as a phonon excitation of the γ band' ", *Phys. Rev. C.*, **51**, 3528 (1995).

56. F. Iachello, N. V. Zamfir, and R. F. Casten, Phase coexistence in transitional nuclei and the interacting-boson model, *Phys. Rev. Lett.*, **81**(6), 1191 (1998).

57. V. Werner, E. Williams, R. J. Casperson, R. F. Casten, C. Scholl, and P. von Brentano, Deformation crossing near the first-order shape-phase transition in $^{152-156}$gd, *Phys. Rev. C.*, **78**, 051303(R) (2008).

58. D. Tonev, A. Dewald, T. Klug, P. Petkov, J. Jolie, A. Fitzler, O. Möller, S. Heinze, P. von Brentano, and R. F. Casten, Transition probabilities in ^{154}gd: Evidence for x(5) critical point symmetry, *Phys. Rev. C.*, **69**, 034334 (2004).

59. P. E. Garrett, Characterization of the β vibration and 0^+_2 states in deformed nuclei, *J. Phys. G: Nucl. Part. Phys.*, **27**, R1 (2001).

60. D. D. Warner and R. F. Casten, Interacting-boson-approximation e2 oporator in deformed nuclei, *Phys. Rev. C.*, **25**(4), 2019 (1982).

61. D. D. Warner and R. F. Casten, Revised formulation of the phenomenological interacting boson approximation, *Phys. Rev. Lett.*, **48**(20), 1385 (1982).

62. R. F. Casten and D. D. Warner, The interacting boson approximation, *Rev. Mod. Phys.*, **60**, 389 (1988).

63. A. Leviatan, J. E. García-Ramos, and P. V. Isacker, Partial dynamical symmetry as a selection criterion for many-body interactions, *Phys. Rev. C.*, **87**, 021302(R) (2013).

64. J. P. Delaroche, M. Girod, J. Libert, H. Goutte, S. Hilaire, S. Péru, N. Pillet, and G. F. Bertsch, Structure of even-even nuclei using a mapped collective hamiltonian and the d1s gogny interaction, *Phys. Rev. C.*, **81**, 014303 (2010).

J. H. Hamilton and Tsinghua University

Sheng-jiang Zhu and Huai-bo Ding

Department of Physics Tsinghua University, Beijing, China

The following remarks were made at the special symposium to celebrate Professor J. H. Hamilton's 50 years of teaching and research at Vanderbilt University in 2008.

1. Introduction

We sincerely congratulate Professor Joe Hamilton's outstanding achievement in teaching, research, and international cooperation. He is one of the world's most well-known physicists who started scientific exchanges with Chinese physicists as early as the beginning of the 70s of the last century. He has made long-term, substantial collaborations with Chinese physicists and many important contributions to Chinese scientific and technological exchanges with the world and to the development of basic and applied nuclear physics research in China.

Our research group has been carrying out international research cooperation with Professor Hamilton's group for more than 20 years.

Prof. Hamilton and Prof. S. J. Zhu at Tsinghua University.

2. Joint PhD Program

Prof. Hamilton, Chinese students and visiting scholars in Vanderbilt University, 1988. From left: S. J. Zhu, H. Xie, J. H. Hamilton, X. W. Zhao, Y. R. Jiang, and W. B. Gao.

In 1984, Professor Hamilton formally suggested and developed a unique joint PhD program with Tsinghua University. Graduate students first take their courses at Tsinghua University, then pass the Vanderbilt PhD-qualified examinations to carry out a dissertation in the US under the guidance of Professor Hamilton, and finally receive a Vanderbilt PhD degree. Through this program, a number of Chinese students grew up and gained experience in various nuclear research frontiers.

3. Adjunct Professor of Tsinghua University

In 1986, Professor Joseph H. Hamilton was appointed as an Adjunct Professor of Tsinghua University.

Prof. Hamilton receiving the certification of Adjunct Professor from Vice-President Zhang Xiao-wen of Tsinghua University in 1986.

Prof. Hamilton giving a talk at Tsinghua University.

4. Cooperative Research

In 1988, Vanderbilt University and Tsinghua began a cooperation to study the high-spin states of neutron-rich nuclei using spontaneous fission. This cooperation has been highly successful. Many new level schemes of neutron-rich nuclei have been identified for the first time. Octuple deformations around the theoretically predicted $Z = 56$, $N = 88$ nuclear region. Chiral doublet bands and the two-phonon gamma vibrational bands of the odd-A nuclei in $A = 100$ nuclear region.

A total of 137 Tsinghua co-author journal papers and 81 conference proceeding papers have been published, among which, more than 35 were published in Chinese journals. Now, we continue to do our research cooperation.

The Tsinghua Group in 2006.

PHYSICAL REVIEW C **74**, 054301 (2006)

Identification of band structures and proposed one- and two-phonon γ-vibrational bands in ^{105}Mo

H. B. Ding,[1] S. J. Zhu,[1,2,*] J. H. Hamilton,[2] A. V. Ramayya,[2] J. K. Hwang,[2] K. Li,[2] Y. X. Luo,[2,3]
J. O. Rasmussen,[3] I. Y. Lee,[3] C. T. Goodin,[2] X. L. Che,[1] Y. J. Chen,[1] and M. L. Li[1]

[1]*Department of Physics, Tsinghua University, Beijing 100084, People's Republic of China*
[2]*Department of Physics, Vanderbilt University, Nashville, Tennessee 37235, USA*
[3]*Lawrence Berkeley National Laboratory, Berkeley, California 94720, USA*

(Received 7 June 2006; revised manuscript received 27 August 2006; published 1 November 2006)

PHYSICAL REVIEW C **73**, 054316 (2006)

Search for octupole correlations in neutron-rich ^{148}Ce nucleus

Y. J. Chen,[1] S. J. Zhu,[1,2,*] J. H. Hamilton,[2] A. V. Ramayya,[2] J. K. Hwang,[2] M. Sakhaee,[1]
Y. X. Luo,[2,3] J. O. Rasmussen,[3] K. Li,[2] I. Y. Lee,[3] X. L. Che,[1] H. B. Ding,[1] and M. L. Li[1]

[1]*Department of Physics, Tsinghua University, Beijing 100084, People's Republic of China*
[2]*Department of Physics, Vanderbilt University, Nashville, Tennessee 37235, USA*
[3]*Lawrence Berkeley National Laboratory, Berkeley, California 94720, USA*

(Received 21 November 2005; revised manuscript received 15 February 2006; published 31 May 2006)

5. Top Prize Award

Due to his great contributions to Chinese nuclear physics, he was awarded the top prize for International Scientific and Technological Cooperation of the People's Republic of China in 2002.

We sincerely congratulate Professor J. H. Hamilton's 50 years of Teaching and Research at Vanderbilt University.

Bird's Nest in Beijing. Welcome to Beijing!!

He has been invited and visited Tsinghua many times. He has given extensive and excellent lectures to introduce the recent progress at the frontiers of nuclear physics including newly introduced theories and methods, and discussed our research cooperation. Meantime, many scientists of Tsinghua University have been invited to Vanderbilt and Oak Ridge to join research works supported by Professor Hamilton.

The Shining Splendor of Science and Humanity in the East–West Scientific Collaboration: Professor J. H. Hamilton and the Nuclear Physicists of Asia

Y. X. Luo

Retired Senior Research Associate,
Department of Physics, Vanderbilt University, USA
Former Guest of Lawrence Berkeley National Laboratory
Retired Professor of Physics, Institute of Modern Physics, CAS,
Lanzhou, PRC
Former Director of the Institute of Modern Physics, CAS, Lanzhou, PRC
Former President of Chinese Nuclear Physics Society

On June 8, 2022, colleagues of the international nuclear structure community held a wonderful Zoom Event, *Science and Humanity — The Extraordinary Life of Professor J. H. Hamilton*, to celebrate Professor J. H. Hamilton's great contributions to research frontiers of nuclear physics and his service at Vanderbilt and worldwide for 64 years. After discussions with major collaborators from Asian countries, I was very happy to talk at the event. Now, it is a great pleasure for me to write this article to include in more detail what I wish to say.

First of all, on behalf of all the collaborators from Asian countries, I wish to express our sincere gratitude to Professor J. H. Hamilton for what he has done for the collaborations and for his support during the past

45 years. Also, I wish to extend congratulations on Professor J.H. Hamilton's great service of 64 years at Vanderbilt and the worldwide community, and his excellent achievements in the research frontiers of nuclear physics. These include his contributions as co-discover of superheavy elements 115, 117, and 118, and especially his key role in the discovery of element 117 and the naming of the element as Tennessine (Ts). It was named after the State of Tennessee based on his strong recommendation. This is noted in a paper "The recommended naming for the elements 113, 115, 117 and 118 and the studies of superheavy nuclei", published in the *Chinese Science Bulletin*, 2016, Vol. 61, 21, p. 2326.

I am sure that all colleagues and collaborators will agree with me when I write as follows: "Motivated by his awareness for the power of collaborations, and by his modesty and belief as well, the great efforts and scientific collaborations headed by Professor J. H. Hamilton are not only productive and successful, but also shining with splendor of science and humanity. With his scientific leading role and humanistic management with thoughtful concern and sympathy, the scientific family of fundamental nuclear research was continuously expanded in the past half century, which is leading in the fundamental nuclear physics research and full of mutual understanding and friendly atmosphere in it." I feel that it is particularly worth recalling the long-term, substantial, fruitful, and helpful collaborations and training programs with Asian nuclear physicists and students developed by Professor J. H. Hamilton. We all are proud of the collaborations because we all were highly involved in the joint efforts and enjoyed very much the great achievements. The collaborations are unforgettable not only because of the great successes in the physics frontiers and Professor J. H. Hamilton's leading role in the research but also due to his moral integrity and great personality shown in the collaborations.

1. A Pioneer in Scientific Exchanges and Collaborations with Asian Physicists since the Beginning of 1970s

The exploring efforts to develop exchanges and collaborations with Asian physicists by Professor J. H. Hamilton can be traced back to half a century ago. Since the very beginning of the 1970s, Professor J. H. Hamilton delivered physics lectures in Japan, Korea, India, Hong Kong, and China, respectively, which won praises in Asian countries for the leading physics, rich information, inspiration, and friendship in the lectures.

As a pioneer in setting up international collaborations with Chinese nuclear physicists, he is the first American physicist who started to bridge the connections between Chinese and American scientists. Since the beginning of the 1970s, he kept contact with Academician Cheng-Zhong Yang, the founder and former Director of the Institute of Modern Physics (IMP), Chinese Academy of Sciences (CAS), Lanzhou, and with Academician Fujia Yang, the former President of Fudan University, Shanghai, and former Director of the Shanghai Institute of Nuclear Research, CAS, China. It is worth mentioning that through a joint effort, a textbook *Modern Atomic and Nuclear Physics* written by Professor J. H. Hamilton and Professor Fujia Yang was published by McGraw Hill and sold worldwide. A second edition was published by World Scientific and is one of their leading textbooks. It is the first textbook written jointly by Chinese and American physicists and widely used in training

Fig. 1.   An early visit by Professor J. H. Hamilton (third from the left) to China, October 4–10, 1978. In the Institute of Modern Physics (IMP), CAS, Lanzhou, he was received by Academician Cheng-Zhong Yang (second from the left), the Director of IMP.

students. In 1979, he took the lead in submitting and publishing a scientific paper in High Energy Physics and Nuclear Physics in Chinese, one of the major journals in China, which was the first paper in this field from the West published in China since 1949. Afterward, as soon as this journal was issued in English also, he and a Chinese scientist jointly published a paper in the first Volume of the English version of this Chinese journal. During this special period of time, he encouraged and recommended Chinese scientists to publish their scientific papers overseas, which they began to do with remarkable success.

Beginning in 1974, just two years after the diplomatic relations were established between China and the US, Professor J. H. Hamilton frequently sent to Chinese physicists new scientific publications and pre-prints reporting the most recent progress in frontiers of basic nuclear physics and the related accelerator technology. On his first visit to China in 1978 and again in 1981, he gave extensive and excellent lectures to introduce, in a systematic way, the progress in frontiers of nuclear physics including newly introduced theories and methods.

At the very beginning of China's opening to the outside world, his elegant lectures and helpful discussions with Chinese physicists brought in valuable rich information about what was going on at the frontiers of fundamental nuclear physics research in the world, which greatly inspired the Chinese scientists with interest, courage and confidence, and were of

Fig. 2.   A visit in October, 1978, by Professor J. H. Hamilton (first from the left) at the Shanghai Institute of Nuclear Research, CAS, received by Professor Xiao-Wu Chen and Professor S. H. Shi *et al.*, Vice Directors of SINR (from the right).

Fig. 3. A photo taken at one of the four elegant lectures given by Professor J.H. Hamilton in the Institute of Modern Physics, CAS, Lanzhou during his first visit to China, October 4–7, 1978. He is talking about the famous UNISOR (see details in Section 5).

great importance for recovering scientific research in China just after the "Cultural Revolution".

The two beautiful lecture series including the new facilities required and new physics studies of nuclei far from stability and nuclear structure at high spin were translated into Chinese and published in the *Chinese Journal Trends of Nuclear Physics* in 1979 and 1982, respectively, and have been widely used for training students and young fellows in laboratories and universities in China for years. In 1978, he played an important part in receiving the first Chinese Delegation of Nuclear Physicists to visit the US, and the visits largely increased contacts between scientists of the two countries.

Many senior physicists of the Institute of Modern Physics (IMP), CAS, Lanzhou, still remember what Professor J. H. Hamilton said at the beginning of his first talk during his first visit there in 1978: "I am very happy to make contributions to the international exchanges of China and the development of the relationship between the US and China, and I have been looking forward to visiting IMP, Lanzhou. I believe that the new development of relationship between the US and China will be one of the most important historical events in the world in the coming twenty years!" We are very happy to have seen that his prediction, as a physicist, has become true! People also still remember the article "New spirit and direction of Science in China" which Professor J. H. Hamilton wrote after his first visit to the country, to introduce nuclear physics research in China to

Fig. 4.   In the 1980s, 1990s, and 2002, Professor J. H. Hamilton (third from the left) visited successively the major Collaborating Group of Tsinghua University headed by Professor S. J. Zhu (in Fig. 5 and Fig. 6, first from the right).

(a)   (b)

Fig. 5.   (a) A visit of Professor J. H. Hamilton (left) to the China Atomic Energy Institute (CIAE), Beijing, received by Professor Z. G. Zhao, Director of CIAE (right). (b) During a visit in the 1970s in CIAE, received by Professor C. L. Jiang (right), Director of CIAE, Professor J. H. Hamilton climbed up the Great Wall after a lecture.

the US scientists' community. Academician Cheng-Zhong Yang translated this article into Chinese and forwarded it to the Chinese nuclear community. Many senior Chinese physicists still remember that after his first visit to China in 1978 and back to the US Professor J. H. Hamilton presented a briefing on "Nuclear Science in China" to the Delegation of Nuclear Scientists of the US who would visit China in the next year, May of 1979. His scientific attainments, modesty, friendship, and concern all remain in the memory of Chinese physicists.

I first knew Professor J. H. Hamilton from his beautiful lectures given in IMP, CAS, Lanzhou in October of 1981, when he presented his second series of lectures in Lanzhou during his second visit there. The lectures were given in more detail and to a wider scope in comparison to the first series of talks he gave in 1978. After his lectures, Professors X. J. Sun, Z. Z. Zhao, and I took the responsibility to translate his four lectures into Chinese. It was an unusual experience for me to make the translation of his talk on Nuclei Far From Stability. The three magnetic tapes with the three-hour talk recorded deeply impressed me and my colleagues with the very interesting physics in frontier, elegant logic expression, and beautiful English. With no change, the translation of his talk itself was found to be a very beautiful review of the frontiers! Since then, I have read many of

Fig. 6.   Reunion by Professor J. H. Hamilton in 2002, Beijing, with Professors S. X. Wen (right) and K. Zhao (left) of CIAE, who worked in Vanderbilt on Kr isotopes and extending the knowledge of shape coexistence in Hg isotopes.

his papers, mainly in the fields of nuclei far from stability and nuclear structure at high spin. Professor J. H. Hamilton's visits to China in 1986, 1988, 1990, 1996, and 2002, successively, have continuously drawn much attention. His lectures, publications, comments, and suggestions have had strong stimulations and are of great help for the major research programs in the institutes, universities, laboratories, and the nuclear physics community of the country.

From the late 1970s on, Professor J. H. Hamilton has made every effort to seek possibilities to invite physicists from the major laboratories and universities of China to join the international collaborations. In 1979, he invited the first visiting group of IMP, China, Professors X. J. Sun and Z. Z. Zhao, to the US. The collaborative research guided by him and the following work by Dr. W. C. Ma and Professor S. X. Wen led to the discoveries of multiplet band structure and a new region of superdeformation with $\beta_2 \sim 0.45$, which was published in *Phys. Rev. Lett.* 1981, cited over 100 times in SCI (see more details in Section 5).

2. Devoted to Help Set Up and Lead the Long Term and Fruitful Collaborations

By continuing to invite and host Chinese physicists to work in his group, Professor J. H. Hamilton has developed various exchanges and collaborations with Chinese scientists over the past 45 years. He and his co-workers in the US, famous physicists Professor A. V. Ramayya and Professor J. O. Rasmussen have paid special attention to focus on the most challenging frontiers of nuclear physics and to create flexible and effective ways for international collaborations with foreign physicists. Numerous formal and informal exchange programs were developed. Over 55 Chinese scientists worked in the US as well as in China to share the experimental data analysis and theoretical interpretations. Through the Joint Institute for Heavy Ion Research, which Professor J. H. Hamilton founded in Oak Ridge and was Director of, a number of nuclear theoreticians from China were brought to Tennessee to do research with scientists there.

Professor J. H. Hamilton has also developed fruitful collaborations with physicists from India and Korea and has been teaching the students

Fig. 7.   A photo of Vanderbilt Group taken together in Berkeley after we talked at the International Nuclear Physics Conference, Berkeley, 2001, full of a cordial and friendly atmosphere in the group.

from the two countries. He has trained 8 young to full professors and 7 graduate students who came from India, of the latter 7, 1 received a master's degree and 6 PhD. One of them came back 18 years later on a Ford Foundation faculty fellowship for a year. They published 20 papers together, including one recently published in *Phys. Rev.* and one in press in *Nuclear Physics*. In total, Professor J. H. Hamilton published 154 papers with Indian scientists. One of the scientists came for about 15 years and published 94 papers together before he went back as a full professor. I wish to emphasize Dr. J. K. Hwang, who is from Korea and guided by Professor J. H. Hamilton, worked at Vanderbilt for 20 years, playing a significant role in the Group, during which they published 369 papers together.

Many Asian visitors and collaborators received praise for their sincere personality, hard work, wisdom, and ability. Chinese physicists have often won the approval of collaborators. While making considerable contributions to the collaborations, Chinese physicists have benefited very much with regard to the theories and experimental techniques concerning fundamental nuclear physics.

(a)   (b)

Fig. 8.   (a) Happy gathering with Professor J. H. Hamilton, Professor A. V. Ramayya, and Professor J. O. Rasmussen, in ICFN5, November 4–10, 2012, and ICFN6, November 6–12, 2016, Sanibel, Florida. (b) Also pictured are Dr. J. K. Hwang, Dr. E. H. Wang, Dr. S. H. Liu, Professor F. R. Xu (Peking University), and me.

For the collaborations headed by Professor J.H. Hamilton, the major collaborating groups and physicists in the US and China are the Vanderbilt Group and Group of Tsinghua University, the latter headed by Professor S. J. Zhu, and physicists from IMP, CAS, Lanzhou; CIAE, Beijing; Peking University, Beijing; Fudan University, Shanghai; Shanghai Jiao Tong University, Shanghai; Huzhou Teachers College, Huzhou, respectively. The Asian collaborators highly involved in the collaborations are S. J. Zhu, J. K. Hwang, Y. X. Luo, E. H. Wang, W. C. Ma, Z. G. Xiao, S. H. Liu, K. Li, X. J. Sun, Z. Z. Zhao, S. X. Wen, K. Zhao, D. Fong, J. K. Deng, Q. H. Lu, X. Q. Zhang, D. Shi, X. W. Zhao, W. B. Gao, H. Xie, F. R. Xu, Y. Sun, J. Meng, Y. S. Chen, G. L. Long, Y. Shi, Y. J. Chen, and Y. X. Liu *et al.*

3. Setting Up the Successful and Unique Joint Student Training Programs

Professor J. H. Hamilton has also established a unique way for training students in both foreign countries (mainly in China) and the US to improve the training for students and to promote the joint research. He is the first to train Chinese graduate students in nuclear physics in the US since 1949. Based on his idea, the first and unique Joint PhD Training Programs with Tsinghua University and Fudan University, training graduates both in the US and China, using the joint data for their theses, were established. In the programs, Chinese graduate students first took their coursework in China. Those who then passed the Vanderbilt PhD qualifying examinations were jointly supported by their home university and Vanderbilt University to carry out dissertation research in the US under the guidance of Professor J.H. Hamilton, and finally received a Vanderbilt PhD degree. During the past 30 years, Professor J. H. Hamilton has provided three sets of high quality, high-efficiency data taken at the advanced multi-γ-detector array Gammasphere in 1993, 1995, and 2000, respectively, to Chinese students and physicists working in China and in the US, which have been proved to be powerful not only in exploring in unknown neutron-rich regions (see details in the following) but also in teaching and training.

Professor J. H. Hamilton was appointed as an Adjunct Professor of Tsinghua University in 1986, and an Honorary Advisory Professor of Fudan University in 1988, respectively. He has personally guided the Chinese students in joint research and effectively brought them to the frontiers and led them to success in the hot and most interesting topics, leading to numerous publications in the journals *Physical Review Letters,*

Fig. 9. In 1988, Professor J. H. Hamilton was appointed as an Honorary Advisory Professor of Fudan University, Shanghai. The President of Fudan University, Professor Xi-Der Xie chaired the Appointing Ceremony. Professor Fujia Yang (left) presented.

Physical Review C, Physics Letters, etc. Since then, 13 students were brought from China in the exchange program; 11 received a PhD, one a Master and one after a year transferred to engineering; 10 more received a PhD and 3 an MS at Tsinghua University.

Through the programs the students grew up and rapidly gained experience in various research frontiers, playing very important roles in international collaboration and, after back to Asian countries, in research in Asian countries (especially in China) as well. Some of the students have become core members in their groups of the countries. One of Professor J. H. Hamilton's former PhD students served as the Vice-Chair of the Physics Department at Tsinghua University, Beijing.

4. Organizing and Hosting Numerous International Conferences and Meetings

In addition, Professor J. H. Hamilton and Professor Ramayya succeeded in organizing and hosting numerous international conferences and meetings which greatly enhanced the scientific exchanges and mutual

Fig. 10.   Taking a photo together with respected Professor J. H. Hamilton in ICFN5, November 4–10, 2012, Sanibel, Florida. Pictured in the photo are Professors Yang Sun, Y. X. Luo, J. H. Hamilton, Jie Meng, Y. H. Zhang, and Dr. P. W. Zhao (from the right).

Fig. 11.   A conference photo of all the participants at the ICFN6, November 6–12, 2016, Sanibel, Florida.

understanding between the scientists from different countries involved in the collaboration. Professor J. H. Hamilton and his colleagues organized and hosted an international conference series *International Conference on Fission and Properties of Neutron-rich Nuclei (ICFN)*, which is oriented to the new frontiers of the nuclear structure of neutron-rich nuclei and of the new modes of fission, and found to be very successful and beneficial to the collaborators involved in the collaboration. He has paid special attention to invite collaborators to be members of the International Advisory Committee of the international conferences, and to support Chinese physicists (over 50 in total) to join the conferences.

5. Important Discoveries and Achievements in the Frontiers of Fundamental Nuclear Physics Made by the Collaborations Headed by Professor J. H. Hamilton

Over the past 45 years, remarkable discoveries and excellent achievements in various frontiers of nuclear physics have been made by world well-known physicist, Professor J. H. Hamilton and the collaborations headed by him. I am attempting to summarize the great successes in the frontiers of nuclear physics as follows, except for the important role he played in the discoveries of superheavy elements.

5.1. *The leading research of nuclei far from stability at UNISOR and the first discovery and interpretation of shape coexistence in complex nucleus*

Professor J. H. Hamilton and his co-workers conceived and led the design and construction of the famous University Isotope Separator at Oak Ridge (UNISOR) and the Joint Institute for Heavy Ion Research. At this advanced UNISOR facility, he became one of the major pioneers world-wide of the studies of nuclei far from stability. Based on the pioneering studies beginning around 1970, including those achieved at UNISOR, studies of nuclei far from stability had then become a major frontier in nuclear science necessitating billion-dollar new accelerator facilities for such studies as constructed or under construction at this time in Japan, the US, France, Germany, Italy, China, and elsewhere. Organized by Professor J.H. Hamilton and Professor A.V. Ramayya, the international collaborations at UNISOR, including the collaboration with Asian physicists, significantly expanded the research programs. Among all the major achievements made at UNISOR, it is particularly worth mentioning the discovery and first interpretation of SHAPE COEXISTENCE in complex nuclei. Based on the extensive study of the low-lying states in Hg and nearby isotopes, Professor J. H. Hamilton proposed for the first time the shape coexistence in complex nuclei. This important discovery was established by many succeeding experiments and theoretical interpretations, and the shape coexistence has turned out to exist also in high spin band structures in vast nuclear regions. The discovery of shape coexistence in complex nuclei is regarded as a milestone for the study of nuclear shape. It changed the paradigm from each nucleus having one fixed shape. This became one of the highlights of the study of nuclear structure both for low and high-lying nuclear states, and so is discussed in advanced textbooks. See the descriptions of this discovery in J. Eisenberg and W. Greiner in

Vol. 1 of *Nuclear Theory*, pp. 280–285 and Fujia Yang and J. H. Hamilton, *Modern Atomic and Nuclear Physics*, pp. 529–541.

5.2. *The discovery of multiplet band structure and superdeformations in A = 70 region, and the physics idea of reinforcing shell gaps, which is one of the major advances of the nuclear shell model*

In the late 70s, the collaboration led by Professor J. H. Hamilton between Chinese physicists (from the Institute of Modern Physics, Tsinghua University, and CIAE) and Vanderbilt University identified the multiplet band structure and superdeformation in the $A = 70$ region. This is regarded as one of the significant progresses at that time in the field of high spin nuclear structure, which led to the discovery of a new region of superdeformed ground states with $\beta_2 \sim 0.4$ and shape coexistence in the light ^{74,76}Kr isotopes, as published in *Physical Review Letters* 1981, cited over 100 times in SCI, and follow up papers in the *Chinese Journal High Energy Physics and Nuclear Physics*, 1982 and *Physical Review* 1982. Subsequently, Professor S. X. Wen from the China Institute of Atomic Energy was brought to Vanderbilt and contributed along with several Chinese graduate students brought to Vanderbilt to the understanding of the Kr nuclei and extending our knowledge of shape coexistence in the Hg nuclei. Based on this Kr work, Professor J. H. Hamilton developed the idea of reinforcing shell gaps for protons and neutrons which can produce superdeformation. The first examples of such were the shell gaps at $Z = 38$ and $N = 38$ and $N = 60, 62$ for $\beta_2 \sim 0.45$. This concept was presented at several international conferences and then published in *J. Phys. G* and *Nuclear Physics Letters* in cooperation with a Chinese theoretician from Lanzhou working at the Joint Institute for Heavy Ion Research in Oak Ridge. The physics of reinforcing shell gaps when protons and neutrons have shell gaps at the same deformation is behind all the regions of superdeformation and stable octupole deformation, subsequently discovered at high and low spins, and is one of the major advances of the nuclear shell model. Now one has important new magic numbers for deformed nuclei leading to superdeformed double magic nuclei in addition to the spherical magic numbers. See *Treatise on Heavy Ion Science*, Vol. 8, ed. Allan Bromley, pp. 38–49. In the 1990s, studies in $A \sim 150$ and $A \sim 190$ mass regions were also fruitful.

5.3. *The breakthroughs made in the studies of nuclear structure at high spin for neutron-rich nuclei*

The study of high spin nuclear structure of neutron-rich nuclei is of great importance for basic understanding of the effective interaction within the nucleus because these nuclei probe a different range of proton and neutron single-particle orbitals. But, unfortunately, there had long been lacking experimental data due mainly to technical difficulties. Since 1988, the collaboration led by Professor J. H. Hamilton with Dr. I.-Y.ang Lee from Oak Ridge National Laboratory (then working in LBNL), Professor J. O. Rasmussen (LBNL), and Professor S.J. Zhu and Dr. Wenchao Ma from Tsinghua University and co-workers pioneered the study of high-spin nuclear structure of neutron-rich nuclei by developing a combination of multi-γ-detector array Gammasphere with a spontaneous fission source. It has been proven that the combination mentioned above and the related powerful and productive triple to four-fold high-quality γ and γ — particle dataset is so far a GOLD MINE. This is a unique effective means to study the high spin nuclear structure of neutron-rich nuclei which have not been accessible and thus nothing known for their structure before the realization of the method. The four-fold matrix γ-γ-γ-γ with 1.7×10^{11} events significantly improved the accuracy of the data analysis, making it possible to reach the previously unknown nuclear region with very low production rate and very weakly populated sidebands. Also, a time-gated triple γ coincidence method (TTGCM) was developed for the first time to measure lifetimes of levels detected at Gammasphere, and the angular correlation measurements were carried out using the triple data to study the multipolarities of the γ transitions.

With the advanced techniques developed, breakthroughs and continuous progress were realized in the previously unknown neutron-rich region: low to high-spin level schemes of as many as 187 neutron-rich isotopes from $Z = 32$ (Ge) to $Z = 66$ (Dy) were established, studied, and published, and significant progresses in the theoretical interpretations were achieved. This breakthrough made in the high-spin nuclear structure of neutron-rich regions was considered a remarkable achievement in the frontiers of nuclear structure.

Octupole deformations around the theoretically predicted $Z = 56$, $N = 88$ nuclear region were identified with first observations and exploring studies of the asymmetric deformations in Mo, Xe, Cs, Ba, La, Ce, and Pr neutron-rich nuclei. The identifications of octupole deformations in ^{143,145}Ba, and then in ^{145,147}La and other nuclei nearby are also the first evidence for the theoretically predicted rotation enhancement of stable octupole deformation in certain nuclei and its quenching in others. One of the papers reporting the progress was cited more than 90 times in SCI. Afterward, octupole deformations were also identified in the very weakly populated light neutron-rich La isotopes ^{143,144}La by our collaborations. The observations and interpretations of two-phonon octupole vibrations in Sr and Zr neutron-rich nuclei, and the first evidence for the coexistence of octupole structures along with β and γ — vibrational states in ^{148}Ce were achieved. Recently, coexistence of reflection asymmetric and symmetric shapes was discovered and interpreted in ^{144}Ba, published in PRL, and it has been highlighted by the editors as Editors' Suggestion. Furthermore, a systematic study in the nuclear region yielded observations and interpretations for sharp drops in electric dipole moments D_0 in Cs, Ba, La, and Ce neutron-rich isotopes, which were accounted for by the reflection-asymmetric mean field shell correction theory.

Remarkable achievements were also made in the studies of nuclear structure in the vicinity of the doubly-magic nucleus ^{132}Sn, mainly in the $N = 83$ and $N = 85$ even-odd isotones, providing valuable information for the check of the spherical shell model. Interesting tilted rotation, or so-called magnetic rotation, was observed in $N = 83$ isotone ^{135}Te by our group which is the first observation of magnetic rotation in the vicinity of ^{132}Sn. The resemblance between the ^{132}Sn and ^{208}Pb regions is also observed. The interplay between single-particle excitations and quasi-collective motions was proposed in $N = 85$ isotones. In addition, the high-spin level scheme of ^{138}Cs was identified for the first time by our group providing important information for the quasi-p-n interactions in this region.

Systematic studies of shapes and structures from $Z = 39$ (Y) through $Z = 45$ (Rh) was carried out by our collaboration which provided experimental and theoretical evidence for the evolution from well-deformed axially-symmetric $Z = 39$ (Y) to near maximum triaxiality in $Z = 43, 45$ (Tc and Rh). Through intensive search in the triaxial region, chiral symmetry breakings were proposed by our collaboration in even-even neutron-rich Mo and Ru isotopes ^{106,108}Mo and ^{110,112}Ru. Professor J. H. Hamilton and the collaborating theoreticians made beautiful interpretations for the doublet bands observed in ^{108,110,112}Ru as the evolution of chirality from γ-softness in ^{108}Ru to more rigid triaxial deformations in ^{110,112}Ru. As a matter of fact, the doublet bands identified by our collaboration in the even-even neutron-rich Mo and Ru isotopes ^{106,108}Mo and ^{110,112}Ru are the best candidates of chiral symmetry breaking so far reported.

Additional important results are briefly mentioned in the following: Systematic studies of disturbed chirality and shape evolutions in even-even Pd isotopes; First observation of wobbling excitations in the first even-even wobbler ^{112}Ru; Triaxial shape evolution in Nb isotopes; First observation of oblate shape in neutron-rich Ag nucleus; Systematic studies of nuclear structure of the long isotopic chain $^{143\text{-}153}$Pr and $^{152\text{-}158}$Pm; Discovery of paring free phenomenon in ^{100}Y and ^{102}Nb; First identification of pseudo-spin doublet bands and Gallenger-Moszkaowski doublet bands in the ^{100}Y; Systematics in band crossings in Tc and Rh isotopes and interpretations; First observation of the yrast band in ^{100}Nb and studies of unexpected sudden shape transition; shape coexistence in Nb isotopes; Observations of shape coexisting structure with different deformations for the $h_{11/2}$ states in Cd, etc.

5.4. *The discovery of cold fission, new insights into fission process and the potential applications to the next generation of nuclear power reactors*

Continuous interest in the studies of fission has been focused on this important nuclear reaction process over the past 70 years. With the development of high-efficiency Gammasphere in conjunction with a spontaneous fission source, a new fission process, very rare fission events called "cold fission", was founded by the international collaboration headed by Professor J. H. Hamilton. The identification of triple nuclear molecule ^{96}Sr–^{10}Be–^{146}Ba was published in PRL, and ternary ^{5}He and scission neutron yield determination and cold alpha ternary fission were also studied. This discovery was regarded by the international community as significant progress made in recent years in the studies of fission process. This process is indeed the ultimate form of cluster radioactivity predicted by Professor Greiner and co-workers in Germany and first confirmed in the ^{14}C emission by ^{223}Ra. As part of this work, the first measurements of the yields of correlated pairs of nuclei formed in fission were carried out for neutron emission from zero up to 10 neutron emissions. Then the energies and intensities associated with each nucleus in these correlated pairs were determined. A new hot fission mode with 7–10 neutron emissions involving hyperdeformed Ba nuclei and then the first evidence for 8–11 neutron emissions in Ce nuclei were discovered. New insights were found into the fission process from the first measurement of the average angular momentum for the primary fission fragments as a function of neutron multiplicity and N/Z ratio, leading to new understandings of how the energy is carried in the descent from the saddle to scission including strong coupling between the bending mode and dipole oscillations. These achievements have drawn increasing attention not only because of the great success in solving the technical difficulties but also due to the theoretical significance of these discoveries. Results were published in the PRL and other journals, reported in international conferences, and are highly esteemed. It is believed that these important discoveries will be included in future textbooks of nuclear physics for updating the knowledge about fission process.

Now, that the collaboration has demonstrated the detailed information that can be measured in spontaneous fission, people have been asked to carry out such measurements of neutron-induced fission of important nuclei like ^{235}U and ^{239}Pu. Such data provide critical input for the design of the next-generation nuclear power reactors. Moreover, these data provide the base for a new method of determining quantitatively the amounts of fissile ^{235}U and ^{239}Pu in spent nuclear fuel and nuclear waste to provide essential information for nuclear waste storage. Thus this work has enormous potential in solving critical problems for society.

5.5. *Numerous publications in international journals and invited talks at international conferences*

The successful and fruitful international collaborations headed by Professor J. H. Hamilton have accomplished numerous publications in international journals, including *Phys. Rev. Letters*, *Phys. Letters*, *Physical Review C*, *Nucl. Phys.*, *Euro. Phys.*, *J. Phys. G* and Chinese journals, among which 233 papers were published with Russian and Chinese physicists; 158 papers were published with just Chinese physicists; 154 papers with Indian physicists; and 369 papers with Korean physicists. Over 80 invited talks were given at international conferences co-authored with Chinese and Korean physicists; The very high esteem of the research work by the international scientific community is strongly evidenced by the remarkable number of invited talks as many as 248 in 37 countries, among which 45 invited talks were given in Asian countries.

6. Beneficial Environment Full of Cordial and Cheerful Atmosphere in the Collaborations

During the past 40 years all the collaborators, students, and visiting scientists were not only satisfied with and proud of our productive and leading research work on the frontiers but also enjoyed very much the cheerful collaborations and cordial and friendly atmosphere in the collaborating groups. Professor J. H. Hamilton has led the collaboration in the research frontiers and personally guided the students in courses and joint research, and has always shown profound concern to the students and collaborators as well. He is warmly praised for his promotion to the friendly relations between scientists from different countries.

It was his scientific leading role and humanistic management with thoughtful concern and sympathy that lead the collaboration to be successful, and the scientific family of nuclear physics was continuously developed in the past half-century. So far, I am deeply moved that many

Fig. 12.   Happy gathering with respected Professor J. H. Hamilton in ICFN6, Nov. 6–12, 2016, Sanibel, Florida. Also pictured are Professors W. C. Ma, F. S. Zhang, S. G. Zhou, Y. X. Luo, F. R. Xu, and Dr. S. F. Zhu.

colleagues still admire the great achievements in research and the mutual understanding and cheerful friendship in our collaboration.

Professor J. H. Hamilton is the founder of the international collaboration, and he is also a good friend to everyone in the collaboration as well. I still often recall my pleasant stay in the Vanderbilt Group. With almost two decades passed, the joint efforts and research achievements of the collaboration are unforgettable and enjoyable, and the fruitful collaboration and friendly relationship remain in my memory! I still remember very well when my father and mother passed away in 2012 and 2014, respectively, Professor J. H. Hamilton and Professor J. O. Rasmussen sent, respectively, e-mails to me to show great concern and comfort. I was deeply moved by their sympathy and thoughtful arrangement. I wish to express my sincere thanks to them again at this special moment.

7. A Great Award Warmly Applauded by the Community: Professor J. H. Hamilton Received The 2002 China Award for International Collaborations of Science and Technology

As detailed above, since the early 1970s, world well-known physicist, Professor J. H. Hamilton pioneered scientific exchanges and collaborations with nuclear physicists and students of Asian countries (mainly of China), then kept continuous long-term, substantial, and fruitful collaborations with the Chinese nuclear physics community. The international collaboration with Chinese scientists yielded a number of remarkable achievements in the studies of nuclei far from stability, of nuclear structure at high spin of neutron-rich nuclei, and of the fission process. These drew much attention and are regarded by the international community as among the major progresses in the relevant frontiers of nuclear physics. Moreover, his continuous collaborations with Chinese nuclear physicists played an important role in promoting the development of fundamental nuclear physics in China. The research program "Synthesis and studies of new nuclides" in IMP and SINR, CAS, China was very successful with 25 new nuclides identified. The studies of nuclear structure at high spin carried out in the country also yielded remarkable achievements. Professor J. H. Hamilton's successful efforts in organizing international conferences and strongly supporting Chinese physicists to join them provided very good chances to bridge the Chinese and international physics communities. The unique Joint PhD Program with Chinese universities developed by him and his colleagues improved the training of students and promoted international exchange and collaboration with China.

Based on all Professor J. H. Hamilton did in the 1970s, 1980s, and 1990s mentioned above, as an outstanding world well-known physicist who is held in great respect in the Chinese nuclear physics community for a long period of time, he was awarded "The 2002 China Award for International Collaborations of Science and Technology". The world's well-known physicists Professor Water Greiner, Professor C. K. Gelbke, and Professor Lee L. Riedinger provided Recommendation Letters for the award. It was an honor for Professor S. J. Zhu, Professor Y. X. Luo, Professor S. X. Wen, Professor W. C. Ma, and Academician FuJia Yang, to submit Recommendation Letters and talks to recommend Professor J. H. Hamilton for the Award.

Since then twenty years have passed. Supported by the National Development and Reform Commission of China, Chinese Academy of Sciences, and NSF of China, research in heavy-ion nuclear physics and applications has realized new breakthroughs in recent years in China with

Fig. 13. Professor J. H. Hamilton received The 2002 China Award for International Collaborations of Science and Technology. The Award was promulgated by the Vice State Premier of the P. R. China.

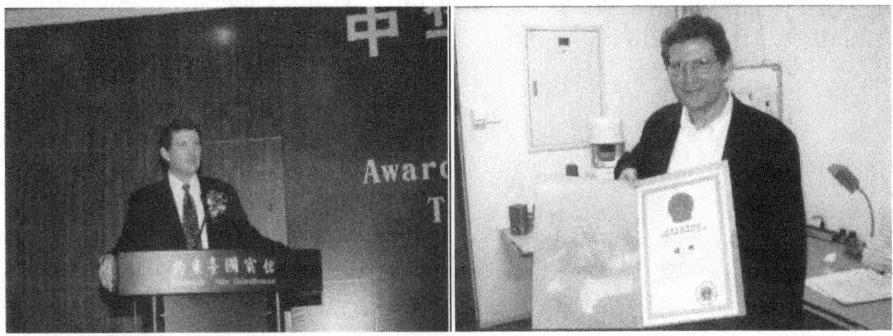

Fig. 14. Professor J. H. Hamilton made a speech in the Awarding Ceremony (left) and afterward showing the certificate of the Award to collaborators (right).

the advent of the Cooler Storage Ring in Lanzhou, Beijing Radioactive Ion Beam Facility, and Jinping Underground Laboratory. With great efforts, over 10 new isotopes are synthesized, a batch of masses of short-lived nuclides are measured with high precision, structure and property of many exotic nuclides are studied, and key reactions in stellar environments are studied. Heavy-ion applications are developed, such as carbon-ion cancer therapy, assessment of the single particle effects of chips and electronics used in space science, and evaluation of materials applicable to advanced nuclear energy systems. The High Intensity Heavy-ion Accelerator Facility (HIAF), a Major National Science and Technology Infrastructure Facility, is under construction and is scheduled to be commissioned in the year 2025, which will bring researchers to the forefront of promoting the most vigorous and fascinating fields in nuclear physics, and open new heavy-ion applications beneficial to societies.

Last but not least, Professor J. H. Hamilton has also devoted many efforts to developing collaborative relations with countries throughout the world. Hundreds of scientists from over 30 countries have been invited by Vanderbilt and the University of Tennessee through the Joint Institute for Heavy Ion Research, for research with them. He has published papers with more than 230 foreign scientists from over 40 countries. He has trained 64 PhD students and over 100 postdoctoral fellows and visiting scholars from over 34 countries during his 64 years at Vanderbilt.

The impact Professor J. H. Hamilton has made on nuclear physics research and collaborations has been far beyond Vanderbilt, and surely throughout the worldwide community!

Joseph H. Hamilton's

William Purcell

Nashville, TN 37235, USA

I ....

There are many sides to Joe Hamilton and, as his nephew, I have been blessed to know the personal as well as the public policy side of his life. Joe Hamilton was born in 1932 in Ferriday, Louisiana, a town of 2,900 people. He was born in a clinic that was located over the drug store on the corner downtown. Ferriday sometimes claims to have more famous people per square mile than any other town in America. But let me report, especially for those of you who are scientists, at less than two square miles, it's a very small town. Mickie Gilley, Jerry Lee Lewis, Jimmy Swaggart, Joe, his older sister, and my mother Mary Hamilton Purcell were all born there. What they have in common is that they all left Ferriday, Louisiana, and went out into the world. Joe's father was Rev J. H. Hamilton, the First Baptist Church of Ferriday pastor for over 20 years. His mother, Letha Hamilton, was a teacher in the public schools of Ferriday, Louisiana. You should know that as a little boy he was in constant motion, just as you knew him later. His teachers would say, "He won't sit in his chair. He climbs up in his chair. He climbs up on top of the desk. He's all over the place." That is what kind of student he was and in church, he was no better. In church, he was so active his mother had to move them both to the back pew because it was so distracting to the congregation for Joe to be down front while his father was preaching.

A particular day that my mother loved to recount was the story of Joe's first "big boy long pants". He was sitting there with his mother in the back of the church when my grandfather began to talk about divine peace. After saying the word peace several times, Joe Hamilton, in his first public address, stood in the back of the church, and proclaimed, "Peace? Peace? How can there be peace when you have to wear wool pants?" This was the first existential question that Joe Hamilton asked in a public forum for which only he knew the answer. In the same spirit, Joe Hamilton came to know how to ask for public support and public funding. Whether it is because he was a preacher's kid and Sunday after Sunday he sat there while his father passed the plate, we will never know. But from an early stage of his career, he understood exactly how to seek and receive the support that he needed to do the work we needed. In slide 1, you'll see him in action in a place I know something about.

Joe Hamilton convinced a group of us to go to the gamma-ray detector at Oak Ridge — a place we had never studied, or fully understood. Somehow, he also convinced the Senate Finance Chairman, Douglas Henry (who in the photo below is standing in the middle with the broadest smile), and a new, young, Majority Leader, me, to see where our money had gone.

The Department of Energy (DOE) had requested proposals for a 20-million-dollar gamma-ray detector, with a one-month deadline. The opportunity to have such a facility would be influenced by how much local funding would be available. Joe first convinced Vanderbilt University and the University of Tennessee to each put up $100,000. He then went to Senator Henry to request $800,000 from the state. Senator Henry called the Commissioner of Finance and Administration, who agreed to have the governor sign a letter pledging this amount. This $1,000,000 convinced the DOE to fund the GAMMASPHERE at Oak Ridge National Laboratory (ORNL). Though this project was ultimately awarded to the University of California Berkeley, Joe was essential to getting the commitment which paved the way for more.

When Joe brought a group of politicians to Oak Ridge, myself included, all of us were thinking, "What are gamma rays, exactly, and why in the world have we come into this place?" The only reason we came, we all knew, was because of Joe. He had arranged for funding from the state and brought that consortium together, bringing people together around things and issues that were important.

Fig. 1.   In this picture, you see a group of mostly political people and that includes President Joe Johnson of the University of Tennessee.

The Joint Institute for Heavy Ion Research was a critical part of the work but first, in the 1970s, Joe obtained money from the state for the University Isotope Separator at Oak Ridge ("UNISOR") project, his first involving state money and funds from multiple universities. The UNISOR project interestingly begins with Joe in a Young Adult group at the First Baptist Church in Nashville where he was sitting with the Governor's Chief of Staff. On that day, in that place, the Chief of Staff became unexpectedly inspired by the work Joe was doing. The result was state funding, and 10 universities came together to support UNISOR.

When ORNL was awarded funds for the first nuclear accelerator to be used by scientists from all over the country and the world, Joe saw the need for an institute to house on-site visiting scientists. The Joint Institute for Heavy Ion Research was initially funded by Vanderbilt University, the University of Tennessee, and ORNL, and subsequently, buildings 2 and 3 included state funds after Joe asked the governor to make the institute a center for higher education for Tennessee. The following January, the

Fig. 2.   Here, Professor Hamilton is lecturing to us all on why this particular instrument at this particular place, that we had provided some particular amount of funding for, was critical to the future of the world.

governor announced his new 20 million dollar per year Centers of Excellence for Universities across the state, including Joe's institute. Before they could move forward, they had to overcome DOE objections to university-funded buildings on DOE property. Finally agreeing, they called it the "Hamilton Hilton". We have the Joint Institute for Heavy Ion Research, again, because Joe obtained those state funds and matched those state funds with contributions from Vanderbilt University and the University of Tennessee, making buildings 1, 2, and 3 possible.

Science advances and new facilities are always needed. Being a visionary, Joe saw the need for a new generation recoil spectrometer (RMS) that could individually separate the multiple products in a heavy-ion reaction. First, he encouraged Vanderbilt University, the University of Tennessee, UNISOR, Idaho National Engineering Laboratory ("INEL"), and Oak Ridge Associated Universities ("ORAU") to each contribute $150,000.

When the DOE decided to fund a much smaller, older, and less expensive RMS, Joe convinced ORNL to provide $200,000 of non-DOE funds, and with Senator Henry's help, convinced the governor to add

$500,000 to get the DOE to provide the balance of funds for the new RMS generator.

When Oak Ridge wanted a new neutron source, they reached out to Vice President Al Gore for a new accelerator but had no response. Then ORNL called Joe to contact Vice President Gore to get his approval. Joe articulated broadly the importance of this new world-class facility for neutron science. He dispatched his nephew, me, to talk to the Vice President. We discussed in detail the importance of the spallation neutron source project. Not many weeks later, not only was the funding approved, but Vice President Gore was in Oak Ridge to announce that we would have a new neutron source. Joe could also play defense. When the DOE was preparing to close the ORNL accelerator, Joe and Lee Riedinger went to Congresswoman Marilyn Lloyd, and the next thing you know, we had a radioactive beam facility at Oak Ridge.

Throughout his career, he's used every public platform possible to push the importance of science innovation in Tennessee. He testified to Congress about the importance of summer undergraduate fellowships at the national labs and they were created. He served for eight years on a Tennessee-DOE committee to study toxic and radioactive releases from ORNL that could have influenced the health of people who lived near the lab. The more than 3,800-page report was the definitive work on this environmental justice issue, which was critical, not just at Oak Ridge but also to the state of Tennessee and other national laboratories. Joe served on the Council of the American Physical Society for eight years and the Board of the American Institute of Physics and the Science and Technology Council for the State of Tennessee for eight years. He served on numerous university, government, and laboratory review committees.

In 1970, when the Air Force decided to close Wright Patterson Air Force base nuclear lab, where Joe had worked, the lab suggested Joe get the Air Force to give the whole lab to one institution. Joe contacted Senator Baker of Tennessee, who convinced the Air Force to agree to transfer the facility to Vanderbilt University. Shortly before the lab was transferred to Vanderbilt, the Triangle Nuclear Lab in North Carolina learned of the opportunity and wanted the Air Force to agree to give the lab to them. By this time, Vanderbilt, the University of Cincinnati, and Duke University each wanted separate parts of the lab, but it was to be given to one lab. Joe contacted Senator Baker's office again. They quickly found the Secretary of the Air Force at an event, and he then directed that each of the three institutions get what they asked. Joe got it done. When Joe wanted to take a small source of ^{238}Pu, a highly reactive plutonium ion, to Canada for an experiment, he discovered he could not because there was a law prohibiting any plutonium from leaving the country. Joe's material was not useful for any weapon and extremely small amount. With Senator Baker's help, the law was changed to allow milligram quantities of ^{238}Pu to be taken to Canada.

The Tennessee Senate and House of Representatives passed a resolution to recognize Joe and his many incredible contributions, not just to Oak Ridge, or the state of Tennessee, but also to science. He has traveled the entire world. He was in Sweden in 1959, and the next thing you know

Fig. 3. Joe is presented with the resolution of the Senate and House by two Tennessee Senators.

Table 1. This summarizes all the fundraising activities in which Joe Hamilton was involved.

Direct Operating Funds 1958–2020	Indirect Operating and Equipment 1974–2021
NSF $ 3,041,000	DOE HRIBF $ 266,000,000
DOE $12,055,000	DOE SNS $2,000,000,000
INEL $ 305,000	State TN $ 16,000,000
UNISOR $12,180,000	Department of Energy J Insts. $3,600,000,000
JIHIR $ 7,710,000	DOE J. Insts. $3,600,000,000
Conf. $ 700,000	**Total $5,882,000,000**
ORNL $ 500,000	
LLNL $ 100,000	TN Centers of Excellence $ 700,000,000
Sum $36,541,000	
Direct buildings and equipment $10,385,000	
DIRECT TOTALS $46,926,000	**Grand Total $6,582,000,000**
Vanderbilt Direct $ 1,037,000	

he's in Russia. Joe was the first American Physicist in Russia since 1938. In 1978, he was in China — the first American physicist there since 1949. From both visits, Joe developed long-term collaborations with Russian and Chinese scientists. He raised $600,000 for the program that was used for the discovery of a new element on the periodic table and was given the honor to name it for the State of Tennessee — tennessine. He also raised funds for many other teaching and lectureships at Vanderbilt, Slack research lectures, and Forman teaching and when his first female PhD student died, her two sons gave $100,000 to endow a lectureship for women in her honor. As a result of his travels, Joe is a 4 million miler on American Airlines.

There is a sense that Joe Hamilton may be the "Forrest Gump" of physics when you look back over the history of his life and career — he was always *there*. He was typically there first, but most importantly, he brought the rest of us along for the journey. Throughout, at the center of his life were his wife, children, grandchildren, and family, together with his undergraduate and graduate students and research associates from all around the world. The results reported here are just snapshots of what Joe's efforts have meant to Vanderbilt, to Tennessee, to Oak Ridge, to the DOE, and to our world. A most extraordinary contribution to the improvement of knowledge and learning, and understanding through collaborations throughout the world he has championed and led.

Direct and Indirect Funds for Research By J. H. Hamilton. These do not include the many millions expended to run accelerators and provide very rare ^{48}Ca for beams.

https://doi.org/10.1142/9789811296963_0026

Joe Hamilton's Contributions to the World of Religion and Science from 1963 to the Present

Floyd A. Craig

Knoxville, TN 37996-1200, USA

1. Introduction

My spouse, Anne, and I have been friends with Joe and Jannelle Hamilton for more than 55 years. We did not meet in a university classroom as you might guess, but in a local church in Nashville, Tennessee, where they were teaching for many years Sunday School classes of sophomore women and college seniors and graduate students about science and religion and how they complement each other. A Vanderbilt Professor of Physics teaching a Sunday School class! How was that possible?

We discovered Joe's religious faith was as important as his diligent, tireless scientific writing and teaching and exhaustive research schedule at Vanderbilt University and Oak Ridge National Laboratory.

When asked about how many people he has taught about science and religion, Joe, who keeps copious notes about everything, estimated the count to exceed 11,000. Our guess is it is likely three times that!

But let me back up. It was almost 60 years ago that Joe's first article was published about learning from the apparent conflicts of science and religion. Without a doubt, in those days, especially in the southern region of the U.S., editors of religious publications took a risk to include such subjects in their publications.

2. Joe was an Early Pioneer in Writing about How Science and Religion should be Understood

As you may recall, at least those of you who are in your fifties or sixties, to talk or write about evolution, creation, the age of the universe, scientific discoveries, physical sciences and how the Bible related to these subjects in a truthful and understandable manner was risky if you wanted to be published. The conventional wisdom then was to avoid this approach to science and religion.

But as it turned out, Joe's articles and early speaking engagements found a small, but growing audience. And with each article, it grew! I believe Joe's non-technical writing and speaking about science and religion were the first for many college students and denominational publications. These set the pattern for the future.

It was almost like there was a hunger for thoughtful non-technical scientific information related to the Bible and to the theology which was being preached and taught during those days. Joe was fortunate to have been blessed with a family who encouraged his search for truth in the world of religion as well as in science. His father was a Louisiana Baptist pastor who took a different view of the conventional wisdom and passed it on to Joe.

During the 1960s, the University of Richmond helped spread Joe's contribution to science and religion when they invited him to present a series of lectures entitled *Preaching in a Scientific Age* to the Virginia Baptist Pastor's School.

Then came a major jump in recognition when Joe wrote articles such as the following:

- "What is Science?";
- "What is the Bible?";
- "Does Science Challenge the Bible?";
- "Faith for the Space Age";
- "The World of Science";
- "The Christian and Evolution";
- "Will Scientific Discoveries Shake Religion?";

Joe was very wise in that he targeted his writing to what college students were asking about science and religion. Astonishingly, all the articles that appeared in student and adult magazines reached a circulation of over 1,000,000 copies. Joe and Jannelle were close collaborators on all these articles.

Then in 1972 came a paperback book written jointly by Joe and Jannelle titled, *Science: Faith and Learning Series*. Then in the 1980s, this book was translated into Chinese. The first half became required reading for Joe's physics course for non-science majors for the past 30 years.

Jannelle joined him again in 1983 to help write the Hatfield Lectures for Georgetown College titled, *Modern Science and Technology*. It is my belief that Joe's early writing on science and religion paved the way for speaking engagements in university and college campuses throughout the south and beyond. To only name a few states, Alabama, Florida, Georgia, Missouri, Virginia, Oklahoma, Louisiana, Kentucky, Arkansas and Mississippi.

A variety of university-endowed lectureships discovered Joe during this period, which again opened the door for more invitations and writing opportunities for these subjects. A fresh and knowledgeable voice had emerged like a cool spring to a dry and parched field. They prepared two special lectures under the heading, *Christians in a Technological Age* entitled, *Science and Technology: Essential But Not Sufficient* and *Christian Realism in a Technological Age*, which were given at many universities.

Perhaps, it was these lectureships and publications which caused the Foreign Mission Board — the worldwide missionary arm of Baptists — to ask Joe to speak with special emphasis on science and religion to universities, colleges and churches in Korea, Hong Kong, and the Philippines Islands on a three-week tour. Then followed speaking engagements on science and religion in Nairobi, Singapore, Bangkok and Tokyo.

In Hong Kong, Joe was the featured speaker at a dinner for over 2,000 Chinese business persons and their spouses. The leader of the program said this was the most significant outreach effort for the Christian faith they had ever had. The front cover of the program is shown in Figure 1.

Fig. 1.   Joe was the featured speaker at a dinner for over 2,000 Chinese business persons and their spouses in a 1970 visit to Hong Kong.

Back at Vanderbilt, Joe team-taught one semester an *Introduction to Theological Education* course for graduate students in the Divinity School. To continue his religious understanding, he took courses at the Divinity School. He observed that he had taken so many courses that had he completed a couple more courses, he could have earned a Master's degree from the Divinity School.

Now let me share with you a few stories Joe related to me. Even after all these years, many of the written and recorded lectures are still available. Some time ago, the Hamiltons received a letter from a biology teacher in a neighboring state. When she was a college student in 1985, she began to struggle with her faith and science. Her mother had kept the 1970 articles on science and religion they had written for adults and gave them to her. These helped her reconcile her problems. Now, over 25 years later, her son was struggling with the same issues and she wanted to give him these same articles she had saved.

But first, she wanted to know: *"Do you still believe what you wrote in the 70s and 80s about science and religion?"* Their response was: *"Absolutely!"* Almost in the same breath, they added: *"We must tell you that the profound scientific discoveries continue to shed light upon our understanding of religion and science."*

Joe was asked to be the banquet speaker at the 50th class reunion of his Mississippi College class of 1954. The college president in introducing him said the following:

When I was a student at Vanderbilt around 1970, I saw that a physics professor from Vanderbilt was to speak at a luncheon at the Baptist Student Center so I went. He chose for the theme of his talk I Peter 3:15 "be ready always to give an answer to every man that asketh a reason for the hope that is in you" His talk had made a lasting impression on him to be remembered after over 30 years.

In 1970, some of his graduate students and Joe held a series of weekly discussions on science and the Christian faith. Over 30 years later, one of the students returned for a visit and said that Joe had changed his whole life. When asked how, he replied: *"As a result of the discussions you led, I became a Christian and that transformed my life."*

Two final activities should be mentioned. Since 1968, they have every month financially supported one to now four children from Korea to Africa to raise them out of poverty, give them birthday and Christmas gifts and write several letters of encouragement a year. Once they bought a house for their child and her family who were about to be evicted. Twice they gave the funds to build churches in Zimbabwe where Joe's cousin was a missionary. These are part of their lifelong giving of 10% and more of their income to their church and missions.

For two years, Joe was the Chair of his church's missions division. He started a program of collecting or buying beds, clothes and food for families in a public housing project. Jannelle followed him as the Chair for two years. They also carried out programs for young people in the project, taking them on camping, summer city park programs and to their homes. One student, who was living with his blind grandmother because his parents had been killed in an accident, wanted to go to college. He took the ACT and made an 8. Joe brought him to his office, made him do ACT exam books and went over his answers every day for three weeks from 9 a.m. to 5 p.m. He took the ACT again and made 18 to be admitted to Belmont University. The President of Belmont called Joe and said he did not believe such a change was possible.

Thank you for the opportunity to throw some light upon Joe and Jannelle's lifetime of service and their amazing contributions to the world of science and religion.

3. About the Author

Floyd and Anne Craig have been friends of the Hamiltons for more than 55 years. Floyd served two four-year terms as the Deputy Director of the Governor's Office of Citizen Affairs and Governor Jim Hunt's Citizen Advocate (Ombudsman) for the State of North Carolina. Prior to his service with the Governor, Craig served as a communications specialist and consultant for state and national denominations. Following this service, Craig Communication Inc. was incorporated as a marketing firm for non-profits, with Floyd as President and Anne Craig as Vice President. Both retired in 2015. Craig has written 10 books on communications and contributed to three more. He is accredited by the Public Relations Society of America.

https://doi.org/10.1142/9789811296963_0027

Local Missions: Furniture, Clothing, and Food

Joe Hamilton

In the 1970s, while Jannelle and I each served terms as the Chair of the Mission's Division, the church was seeking to develop direct outreaches to our community. One such program was to provide furniture, clothing, and food for needy families in the Tony Sudekum Public Housing Project in the area of Carroll Street. This program led to long-term support for several families there.

George O'Neal provided the truck, and together, he and I, with others on occasion, made furniture deliveries to many needy families. The needs of the families were identified through the social worker, who served that community. We made furniture requests to Glendale and to our friends, and we received beds, sofas, chests, stoves, refrigerators, other household items, and clothing, which were delivered to needy families. When specific requests for particular items were not available, we made the rounds of used furniture stores on Nolensville Road, bargaining on price to purchase these with Glendale mission money. At Christmas, we delivered toys and food to a number of families with special needs. Through Don Ramage and Mike Awalt, we learned that Belmont University was replacing dormitory beds and mattresses. Glendale bought about 20 beds at $10 per bed and mattress from Belmont, and we delivered them. A few specific stories will illustrate our activities more clearly.

In one family, the mother was dead, and the father was caring for his nine children, ages from about 4 to 15. They had a sofa (not a sofa bed) and one bed in their two-bedroom apartment where they all lived. So, most slept on the floor. One Saturday, we came to deliver three beds around 9:30 a.m. Three elementary-age children were still sleeping together on the sofa. They were really excited to have these additional beds.

One elderly widowed lady, Mrs. Conklin, was living alone in an apartment on less than $100 per month from her husband's pension. She had no stove or refrigerator. She heated her food by placing cans in pans of hot water from the sink. Glendale provided her with a stove and a refrigerator, for which she was deeply grateful. Then we asked her if she got food stamps to supplement her income. "No" was the answer because you had to go downtown to fill out forms for food stamps and she was crippled and confined to a wheelchair. She had not been out of her second-floor apartment in six years. Martha McClendon knew people who administered the food stamp program, and she got a worker to go to Mrs. Conklin's apartment and certify her for food stamps. The food stamps, stove, and refrigerator significantly improved her living conditions. Jannelle, our children, and I continued to visit with her. One of the most deeply moving experiences was one day when Melissa and Chris were with me, and Mrs. Conklin offered me five dollars to help pay for my gas. I refused, saying we were happy to help her. She replied: "But you have two children you have to put through college someday, so you need to take the money." I explained Vanderbilt would help with college. We continued to visit and bring food and clothes to her until she died. In her very sparsely decorated apartment, one item stood out: the Purple Heart Medal for her son being killed in action in World War II.

Another family we continued to help over 10 years had a crippled son, James, about 20. We took clothes and food to them several times a year. On every visit, James would crawl to the landing at the head of the stairs and call my name. Every time he would say, "Joe, I want to go to school." During that time, our building housed a school (EDC) for children with special needs. Part of our agreement was that we could have one student in the school for free (I was thinking of James). They tested James, but unfortunately, they felt he was not suited for even this school. I was sick. James never failed to recognize my voice nor to ask about school. It was special fun at Christmas to bring gifts to all the children and grandchildren in the family.

Another family we helped over a number of years with clothes, food, and encouragement consisted of a blind grandmother and two teenage children. Their parents had died in a car crash. The boy wanted companionship and help. I took him to Vanderbilt to games and helped him through a serious accident. When his blind grandmother died, I spoke at her funeral. He flunked out the spring of his senior year when she died. I encouraged him to return to school to follow his grandmother's desire for him to graduate from high school. I attended his graduation and encouraged him to think about college, mentioning TSU. However, he wanted to go to Belmont, but his ACT scores were much too low. So, I coached him for three weeks going through the ACT preparation book. He doubled his ACT score and attended Belmont. In my last contact with him, both he and his wife, whom he met at Belmont, were managers of separate Pizza Hut restaurants.

These are the most memorable of the many families we helped in numerous ways. Our clothes room at Glendale grew out of that program, we people brought clothes for us to distribute. One of the fun times was to give a young woman in the housing project a nice, genuine leopard skin coat given by a friend of a church member in a big box of clothes. Our church, through gifts of furniture, clothes, and food, touched and changed many lives.

A final happy time. I was asked to provide transportation for four middle school students in the housing project for an 8-week city summer camp program 5 days a week. My wife, I, and three of my graduate students did the transportation, taking one day a week. We gave each student a candy bar when we took them home. One student was autistic and could not speak. I still remember the great excitement he had when he discovered the Mounds candy wrapper had a second candy in it and his excitement in letting the other students know about this. He lived with his grandmother and one day she came to meet us. She said she did not have anything to give to thank us for doing this, but she said I take in laundry to iron and I would be happy to iron any clothes for Jannell for free. It is amazing how people in the housing project want to do things for you. Such exciting times for us too.

This and the following article by Jannelle were written for the 50-year history of Glendale Baptist Church in 2001.

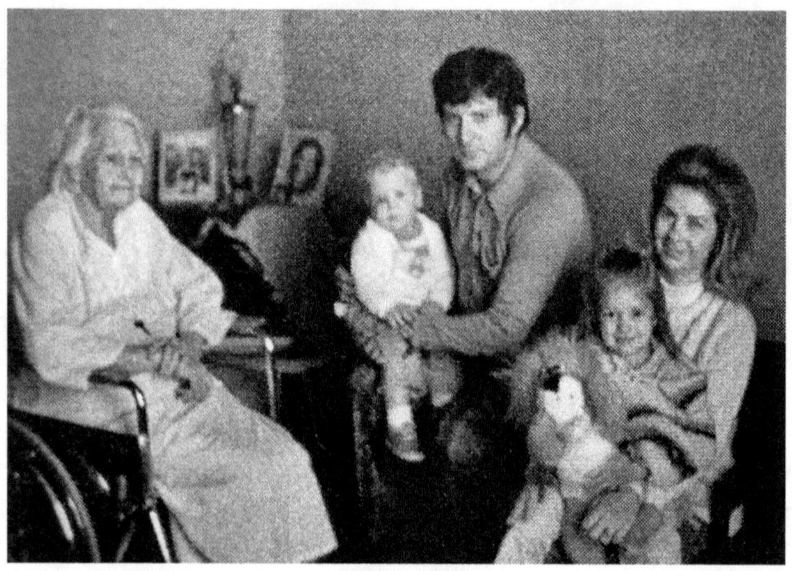

Joe and Jannelle with kids, Melissa and Chris, visiting Mrs. Conklin in her apartment.

https://doi.org/10.1142/9789811296963_0028

Graduate Services

Jannelle Hamilton

In the late 1970s or early 1980s, the church began to feel a need for a service for our young people who were graduating from high school. We viewed this as a very different service from the typical service many churches do to honor graduates. Often, we would only see these young people sporadically after that, perhaps on holidays, so this was a serious goodbye in many cases. Just as their parents were seeing this as a strategic time in their relationship with their children, we saw this as a sort of commissioning and a send-off to the rest of their lives.

Many of our youth had grown up together in this church. We had watched them grow; we had provided Christian education for them; we had loved them as an extended family. We wanted to have a service that would show the great value in which we placed them. We wanted to send them out as important members of our church family, knowing that this body of people loved them and would continue to love and pray for them.

We devised a service of worship around thanking God for them and sending them out with our love and God's presence in their lives. The graduates would each have time to speak briefly to the church. Often, they thought about this for a long time in anticipation of what they would say when it was their turn to graduate. I know that it was a service of great importance to my son and daughter. It was often their only time to speak to the congregation on their own. Then a person in the congregation, who was chosen by the graduate, spoke about this young person. Selecting that person and hearing what this member had to say, especially about them and to them, was very important.

I always felt that those services were like what the early church must have felt when it sent out its own to carry the Good News. We were sending out our most precious ones, and we were asking God to bless them and keep them, and that His countenance would shine upon them.

A Fun Story Told to Joe Hamilton by One of His Church Friends

One of my Church friends got on a plane in New York to fly back home. Sitting next to him was Amand Faessler, a Distinguished Physics Professor from the University of Tubingen in Germany, who was coming to see me in Nashville. They started talking and discovered they both knew me. My Church friend asked Amand if I did much work in physics because he said I did so much church work he didn't see that I had much time for physics. To which Amand replied: "He does so much work in physics I don't see how he has any time for Church work!"

Joe Hamilton and Amand Faessler together at a conference on Sanibel Island in Florida organized by Joe with Amand as one of the invited speakers.

J. H. Hamilton's Remarks at the End of the Zoom Meeting

This Zoom has been an unbelievable, fantastic time. I am truly overwhelmed by the depth and breadth of all the talks. God has been exceedingly good to me and my wife throughout our lives in so many ways. My wife, children, and grandchildren are the major blessings of my life. I much appreciate the outpouring of friendship from each of you in the Zoom. My life has also been enriched by the family-like relationships I have had continuously with all my graduate students, research collaborators, especially Ramayya, and close friends.

For my family and I to become close friends and associates with people from every continent in the world and to experience so many different countries and cultures have been highlights of our lives. Travel has extended from the Himalayan Mountains of Tibet and all across China, to so many areas of Asia, to every country in Europe, many in Africa, Central and South America, the islands of the Pacific, and the Caribbean to make our lives so full and rich. Growing up in the great depression and WW II in a small town in northeast Louisiana with less than 3,000 people, I never dreamed that my life would unfold in the many, exciting, different ways it has, especially with my soulmate wife. When I went away to college, I had gone once to Texas and a number of times to Mississippi (essentially Natchez). My dreams about the future ran to being a baseball player, cowboy, or opera singer. Having played football and baseball in high school and college, I twice went to the LA Dodgers fantasy baseball camps. At the first camp, I drove in the winning run against the Dodger greats for my team when Maury Wills missed throwing his glove at my hit and in the second I won in one game the outstanding defense player in the game award and my team won first place team trophy. I was invited to go with a friend to a ranch in southern New Mexico where we rounded up 500 head of cattle and branded their caves. My wife said when I called her the first night, I sounded like a kid let loose in a candy store. Finally, I sang in the Mississippi College concert choir as a first tenor in 47 concerts to earn a varsity letter in music. So, it wasn't all dreaming.

I cannot thank each of you enough including my family who have spoken so meaningfully and to express my gratitude to you and to all of you who have watched on Zoom. No one could ask for more affirmation of their life than what I have experienced today.

Finally, I want to thank Da Hsuan Feng, Lee Riedinger, and Mark Riley for all their hard work in organizing this event and pulling it off. It has truly been fantastic. I also am deeply indebted to them for arranging to publish all of these talks in a book.

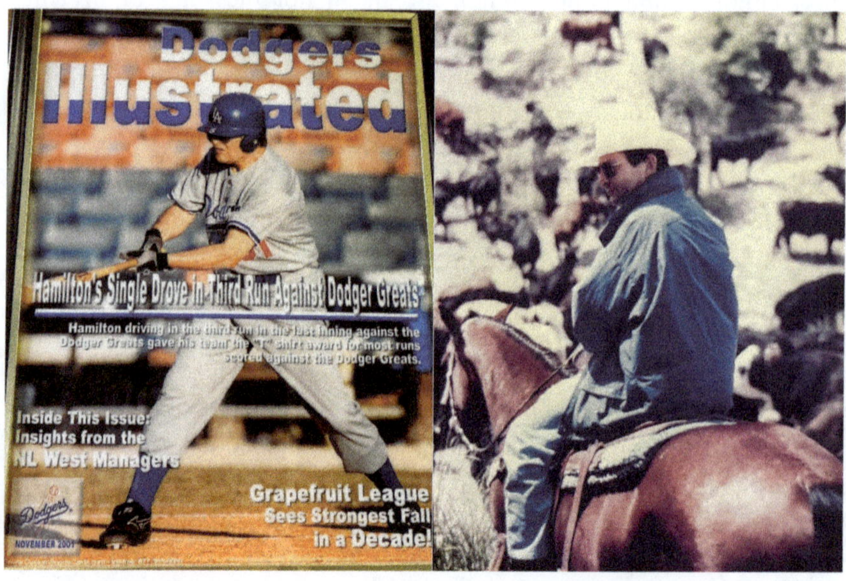

For the book, I thought it would be interesting to describe in some more detail my life and career. About two years ago, I wrote such a description which is attached next.

But most importantly, God led me to meet the most wonderful and lovely young lady the summer before I went to Sweden. We wrote while she was in her senior year at MC. When she graduated, she was offered positions in Nashville and Richmond. She accepted the position in Nashville and the next summer we were married. God truly completed my life in the most wonderful ways through my wife Jannelle and later with two children and three grandchildren.

https://doi.org/10.1142/9789811296963_0031

The Love of My Life

Jannelle Hamilton

Nashville, TN, USA

For a good while in my college days, I had prayed that I would meet the person that God wanted me to marry. On two occasions, I thought that had happened, but as time went on I realized that was not the case. The summer after my Junior year, I was working for a staff member at the college I was attending. My employer had been in school with Joe, and he introduced us. Joe and I had a few dates before he left for a year's work in research in Sweden. During that year, we wrote long letters to each other. For those of you who grew up in the computer age, we did not have access to e-mail or texts. So, we wrote letters on paper and mailed them at the Post Office. We learned a lot about each other that way. We also learned that he had accepted a job teaching and doing research at Vanderbilt University in Nashville, TN, when he returned and that I had been offered jobs in Nashville and Richmond, VA. I later accepted the job in Nashville and left for Nashville the day after I graduated from college.

In Nashville, we began dating again. A year later we married. Two years after that, we spent a semester in Amsterdam while Joe did research at a scientific facility there. It was the worst winter there in eighty years. The Zuider Zee froze over and cars drove over it to two islands that were fishing villages. For a girl from the deep South, that was quite a shock. I knew I was in trouble when I began wearing my winter coat in August. But it was a very interesting country to visit, and we took advantage of our time there to see as much as we could. Before Amsterdam, we bought a car in Paris and toured twelve countries in three weeks; it was a wonderful time.

When we returned to Nashville, we began looking at houses. We loved the hills around the city so we began looking for a lot on a hill where we would have a great view. We finally found just what we wanted and bought it. Then we began to plan our house and designed it ourselves. We built a house with large windows and a balcony so we could enjoy the view. We have loved the house and the view. And because of some woods around us, we experience lots of wildlife such as many deer, a fox family, our own weather forecaster, a groundhog, an armadillo, wild turkeys and their babies, and four raccoons who had been chased into the top of a tree by our dog. We feel so fortunate to live here.

We traveled a lot to physics conferences all over the world. I counted over sixty that I got to visit for pleasure and for the conference.

Our Christian faith has always been a large part of our lives. Over the years, we were members of two Baptist churches, and now we are members of a Presbyterian Church. The Baptist Publishing House asked us to write some materials for adults on the subject of Science and Religion. Later they asked for a booklet for college students. We found that it is pretty difficult for a couple to write together. It causes a lot of *spirited discussions*.

We were so thrilled when a baby daughter came into our lives and later a son. They brought us such joy and enriched our lives in many ways! Now, three grandchildren give us such joy!

I am so proud of Joe's awards and recognition. He has received eight honorary doctor's degrees and so many awards for both teaching and research! After 64 years at Vanderbilt, he has finally retired. We are so proud of the graduate students who worked with Joe. We consider them family. We have a file with a *bucket list* of things we want to do and see in our time left. The first one was a trip to Antarctica which was incredible. We both look forward to other great trips and experiences.

My Amazing Dad:
Joseph H. Hamilton

Melissa Dashiff

Nashville, TN 37235, USA

The daughter, Melissa, of Joseph H. Hamilton reflects on growing up with her famous father and recounts a few of her most memorable moments with him.

Everyone knows Dad is a brilliant nuclear physicist. But he is also a wonderful father whose zest for life helped to shape my life.

One of my first memories with my dad is riding horses in the park. I must have been about five years old. I was riding Comanche, a horse I thought was "mine". The trail was peaceful and sunny and Dad kept turning around to make sure I was ok. Dad loves horses and cowboys — when he was a child, he wanted to grow up to be a cowboy, or a baseball player, or an opera singer. I'm sure he didn't know he'd grow up to become a world-renowned nuclear physicist.

When I was young I loved the times when Dad played records in our den. I would stand on the brick hearth and pretend to sing all the words to his favorite songs: *Rhinestone Cowboy* by Glen Campbell, *You Light Up My Life* by Debby Boone, *Delta Dawn* by Tanya Tucker, and *The Devil Went Down to Georgia* by Charlie Daniels. They became my favorite songs too. I played *You Light Up My Life* on the piano for the school talent show. Dad loves a wide variety of music from *Nat King Cole to Elvis*.

He had friends who were backup singers at *The Ryman and The Grand Ole Opry*. I have been lucky enough to go backstage with him a few times.

There are beautiful woods all around Mom and Dad's house. But when I was a child they seemed like a forest where I could play and explore for hours. Dad built a tree house for me in those woods which I loved playing in as a child and reading in it as a teenager. The tree house was made so well that my own children got to play in it when they were young.

Christmas Eve Gift! Dad shouts every Christmas Eve. It is one of my favorite traditions even though I'm not exactly sure what it means. Dad says his mother would announce it every year. Christmas was always the best time of year at our house! Mom constantly baked delicious Christmas breads and cookies and Dad always had Perry Como playing on the record player.

One day during middle school I was taking a test and looked out the window to see Dad running through the parking lot. Of course, I thought something was wrong but no — there was a solar eclipse and Dad wanted me to see it! I don't remember what the test was on but I'll never forget standing in the school parking lot with Dad watching the eclipse.

Fig. 1.   I have always loved horses.

Fig. 2. In China.

Some of my favorite memories are from an incredible trip with Dad to China. For almost a month we traveled to Shanghai, Beijing, and Xian visiting temples, parks, universities, and The Great Wall. One of the best meals of my entire life was at a dumpling restaurant in Shanghai. Dad and I must have eaten at least 100 dumplings! It was a life-changing trip and I'm so lucky that I got to go with him.

I feel blessed that he is my father. Our whole family is lucky to have him in our lives. He is a kind and supportive father-in-law and grandfather. He has always treated my husband, Duncan, like his own. And our children, Stephen and Adelaide, are lucky to have such a loving grandfather.

My Dad

Chris Hamilton

Over the years, I have tried reading publications by Dr. Joseph Hamilton. I don't understand what I'm reading. I will try to read a sentence and look up words I don't know, but even after reading the definitions, I don't quite get it. Heavy ion? Heavy element? Island of decay? Huh? One time, I sat down with the sole purpose of spending however long it took to understand one of his 1,200 published papers. I got through one page. I just don't get it, but I do know how important it is to him. I also know he is incredible at what he does and he loves physics! He approaches physics and life with purpose and truly enjoys the journey. I have a tremendous respect and admiration for him. I love him, he's my Dad.

When I started playing football in elementary school, he used to tell me about the extra running he did after football practices in the sizzling humidity of a little town in Louisiana named Ferriday. He can prove it to you too because he will let you feel his thigh and/or calve muscle anytime and they are still rock hard. He is 5'10" and weighed around 170 during his college career, where he played both baseball and football. By his junior year at Mississippi College, he was the starting center and place-kicker. Sports, especially baseball and football, were big for him growing up. He also plays golf, tennis, basketball, fishes, and shoots pool. Seeing that side of him as I was growing up normalized him to me because I didn't understand science, especially physics. I liked that he was good at sports *and* physics. The two often don't go together. Scientists are often stereotyped as not athletic.

I have fond memories of him spending hours playing sports of all kinds with me. We just needed a ball. I also loved going to Vanderbilt games with him. Basketball games were exciting at Memorial Gym and he taught numerous players on the men's and women's basketball teams. Besides basketball players, he has taught some 10,000 undergrad students, including Vanderbilt bowlers, tennis players, golfers, football and baseball players. Vanderbilt baseball posters are all over the walls of the hallway leading to his office, with autographs of course. Currently, Vanderbilt baseball is one of the top programs in the nation and it delights him. He grew up a Dodger fan and I get frequent updates on former student and current Dodger, Walker Buehler. At all times, he knows Buehler's record and ERA. He listed stat after stat to me recently about another former player from 20 years ago, Warner Jones, and where he placed in the all time record books for Vanderbilt and the SEC. Former Fed Ex Champion Brandt Snedeker was at Vanderbilt's homecoming recently. My Dad tells me what tournament he's in each week and what his score is on Thursday, then Friday, then Saturday, and then Sunday. Dr. Hamilton cares about all his former students. He talks about them all with such pride.

Hearing that he has taught at least 40 NFL players, including 5 Super Bowl Champions, physics, always stood out to me because Vanderbilt football is a little different. I tell people that being a Vanderbilt football fan builds character. Losing is the norm, but he always comes up with reasons on Sunday why Vanderbilt could have won Saturday. Vanderbilt could be down 42–7 in the 4th quarter, having won no games and he will stay because he loves Vanderbilt sports and is an eternal optimist. He is the most optimistic person I have ever known, almost to a fault. Optimism and perseverance have served him well though. I don't know if he realizes how optimistic he is, but he knows he's persistent. "It can't hurt to ask", I've heard him say many times.

It wasn't just Science and Sports as I grew up though. He has released numerous religious publications with his beautiful wife, my mom. She is a genius by the way also. Science, Politics, and Religion were hot topics in the Hamilton household growing up. I never understood what the hubbub was around Science and Religion but in the southern region of the United States, there sure is. His religious views were unconventional, but not to shock and disturb. Tennessee was home to the Scopes Monkey Trial only 100 years ago, so his views were met with skepticism and anger. I would hear that Science and God didn't mix when I was growing up, but it is the opposite. I noticed how intrigued people were about his stance on evolution and it made me proud that he spoke with conviction about what he believes, and knows to be true.

Part of being a man of God is displaying a willingness to serve others. He is a willing servant, as is my mom. They go the extra mile and do things for people that others won't. As I was growing up, I watched him serve people in low-income housing in a variety of ways. He brought food, furniture, and gifts for people in need. I clearly remember seeing people struggling to live in the times I accompanied him. That had an impact on me. From working with people in low-income housing, he found out about kids who needed rides to camp one summer. He *asked* his graduate students to take them and told them to bring the kids a candy bar every time, probably an almond joy. Sometimes I would be with him when he was tutoring a young man who was living with his grandmother because his parents had been killed. I didn't know it at the time, but years later found out he had made an 8 on his ACT. The two of them worked every day for the next 3 weeks and he scored an 18 the next time he took the ACT. The man was then accepted to Belmont University. It is an incredible feat to help someone raise their ACT score by 10 points in 3 weeks. But it is a far more incredible feat when you think of what the world would offer someone who could not get into college vs. being a college graduate. His potential for a better quality of life increased dramatically after those 3 weeks.

I grew up with Compassion International literature and pictures on the refrigerator. From 1968 to the present, he has supported a child each year from Korea and Africa and even bought a house for a family in Indonesia. He currently supports 3 children, yearly. Family and Community are important and both parents are outspoken about their beliefs and expressing them. They have great affection for human beings. All this played a role in my life and I have been able to serve many, many people because of my parents through God. Observing the way my parents treated others was important to see for me. I have always felt the need to help or serve others and it started with my parents, the love they showed me, and the love they show for others.

I could not have possibly been able to develop a quality relationship with God if my basic childhood needs weren't being met and I liken the parent/child relationship to our relationship with God. In googling "basic needs of a child", listed were, food, water, shelter, education, consistency, structure, and guidance. Check on all those. It was all offered for me at home. I always felt safe and loved at home by my parents. He has lived in the same house, worked at the same University, and been married to the same woman for over 60 years. How's that for consistency?

The thing that stands out to me, maybe more than anything about him, is his comfortability level with God. It's like he doesn't question anything regarding God. He doesn't seem to struggle with the same questions many of us have about God or the afterlife. I truly believe that his only fear of dying is dying before 100. He knew while he was driving many years ago when his father died, almost the exact time and it freaked me out slightly. He said he felt it. I heard of people experiencing things like that but didn't actually believe it until that happened. That and him being sandwiched between a bed of nails, and wearing a turban, made me wonder where in the hell this man came from.

A really cool thing about being Joe Hamilton's son is that I get special treatment sometimes. A few months back, I needed work done on my car. The car place I go to is only because they know Joe Hamilton and I know I will be treated fairly there. On this particular day, the lady I was speaking with said they could not take my car until the next week. I said ok and made time to bring the car back. The owner of the store came out, as I was

about to leave, and asked what I needed. He told me they could take the car right then and have it fixed within 2 hours. The owner told her that I was Dr. Hamilton's son.

When I went to doctor appointments growing up, the physicians were familiar with him and that's always a plus. They would ask him about his work and were obviously impressed. I liked that. Hotels were the same way, airports were the same way, and American Airlines loves him still. It's pretty cool to get bumped up to first class, just because. There is a difference between the way everybody else gets treated and the Joe Hamilton treatment, as I call it. But it's not a money thing as it is with so many people who get better treatment than the rest of us. It's a respect thing 80% of the time and the other 20% of the time is because he raises his voice and "perseveres", so to speak, until he gets his way. I was used to that growing up and it created a difficult dynamic for me as an adult. I know the difference between being treated just okay versus getting the Joe Hamilton treatment.

There are endless examples of things I've gotten to do that others don't get to do. It used to embarrass me, but now I'm thankful for it and appreciate it. In 1990, the Chicago Bulls played an exhibition game at Vanderbilt and I was in the locker room to get to meet Michael Jordan afterwards. That was a really big deal for me. My mom got to see Robert Redford in person because he had the drop on where Mr. Redford would be eating that night while filming in Nashville. I've gotten to travel and have had access to people, places and things I would have not otherwise had access to. It's been very special to go to China, Africa, throughout Europe, and many, many other places. I have a better understanding of the world because of it. I've even been in National Geographic touching a silver sphere with my hair going all over the place. I just shake my head or shrug my shoulders because I know how lucky I am to be along for this ride of his.

In his free time, you can find him looking for hard-to-find coins, patches, and stamps. He has a collection that any collector would be proud of and it makes it easy for me to shop for presents. When he gets presents, he still gets genuinely excited and appreciates any gift. It's really refreshing to watch the appreciation both my parents have for gifts and it says something about their character, I think. This man will still say "ooooh" at the site of a good dessert or watermelon and use his finger to clean the plate if the sauce was good enough and no bread is available.

While on vacation, he reads old westerns. I don't know why it's only on vacations, nor did I understand why he would stay up all night reading them in the bathroom at the hotel. He'd finish a book a night. It's crazy how fast he reads, sitting on a toilet, in a hotel bathroom, with the light on all night, while the rest of us just want to sleep. It annoyed me as a child, but I'm smiling as I write this now. He still appreciates the small stuff and there is a simplicity and naturalness to him that is difficult to describe.

I love his office at Vanderbilt. When I was young, I especially liked going. He kept saltine crackers in a drawer and had a small refrigerator with cokes in glass bottles. His chair was big and comfortable and I could spin around in circles. There were gadgets, articles, books, pictures, plants, and some awards. He loves plants and still has a plant from his mom in his office. There were large bookshelves with glass doors that had many more books I cannot understand. There was a couch that I loved to sleep on. It is so comfortable. When I was 21 or 22, I was trying to prove I could live on my own after quitting college to work in television. I used to drive over to Vanderbilt to spend the night on that couch because I had no air conditioning. I remember laying there, just feeling at peace, thinking "this is wonderful". It is a beautiful, warm office and I truly like being there with him.

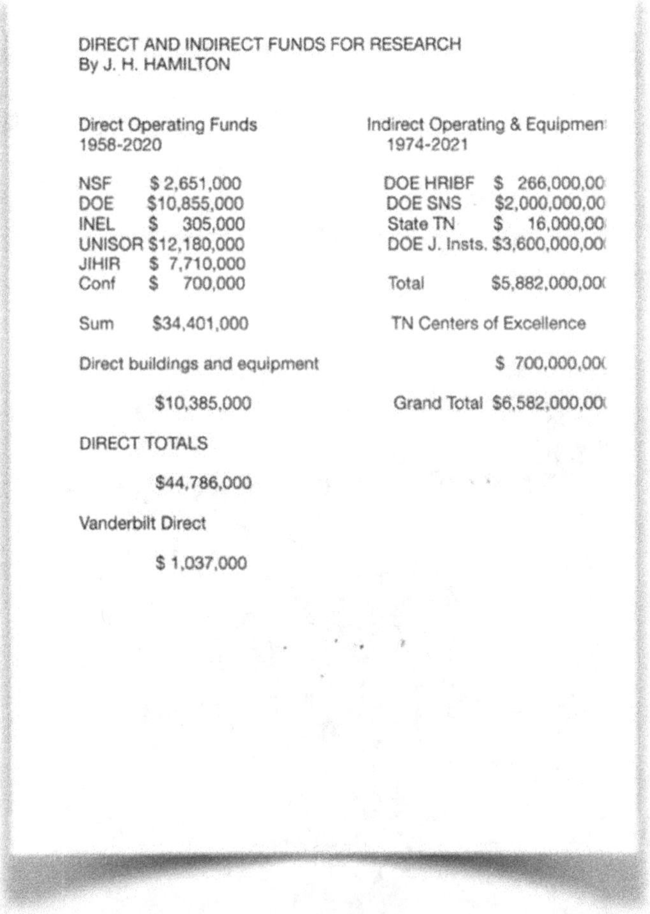

DIRECT AND INDIRECT FUNDS FOR RESEARCH
By J. H. HAMILTON

Direct Operating Funds
1958-2020

NSF	$ 2,651,000
DOE	$10,855,000
INEL	$ 305,000
UNISOR	$12,180,000
JIHIR	$ 7,710,000
Conf	$ 700,000
Sum	$34,401,000

Direct buildings and equipment

$10,385,000

DIRECT TOTALS

$44,786,000

Vanderbilt Direct

$ 1,037,000

Indirect Operating & Equipment
1974-2021

DOE HRIBF	$ 266,000,00
DOE SNS	$2,000,000,00
State TN	$ 16,000,00
DOE J. Insts.	$3,600,000,00
Total	$5,882,000,00

TN Centers of Excellence

$ 700,000,00

Grand Total $6,582,000,00

Almost everything in his office is still the same, minus the crackers and cokes. The big difference is that the walls are completely filled with awards and honorary degrees. The amount of awards he has won is incredible. At some point in the 2000s, he was getting so many awards and recognition that it became old hat. One day when he told me he was getting recognized for an achievement, I said okay and went about my day. Later on, it hit me that I didn't even congratulate him. I was just that used to hearing about awards and honorary degrees. It was basically like him telling me he had to go to a regular check-up at the doctor's office. That really made me realize how incredible, all his achievements are. He has made major, valuable contributions to the world and few can claim that achievement. He would probably argue that the amount of quality, successful PhD students he has taught and/or mentored in his likeness if you will, makes his total contribution immeasurable. They have gone out and made great contributions in their own right. Even more, mentioning his contributions to the world through Physics doesn't do him justice. He has made countless contributions in the world religiously and socially. He just told me recently about a man who told him that his views on God were shaped by him and how thankful he was for Joe Hamilton.

When I was younger, I remember reading about self-actualization from Abraham Maslow. I didn't know and still don't know if it's even possible to truly be a fully, self-actualized human being. Obviously, it's not something that can be measured. But I can say, with confidence, that Joe Hamilton is as close to realizing his full human potential as anyone I have ever known. He has far exceeded any expectations for a small boy, from a little town in Louisiana. I am so fortunate he is my Dad and that I get to be a part of his continued journey through life.

My Life and Career

J. H. Hamilton

*Department of Physics and Astronomy,
Vanderbilt University, Nashville, TN 37235, USA*

A brief history from my birth through high school and universities on to my career as a professor and experimental nuclear physicist including the most significant scientific discoveries is described. The importance of my many international collaborations and work with graduate students, research associates, and long-time collaborations with Professor Ramayya and others is presented. The special importance in my life of my wife, children, and grandchildren is emphasized. Finally, my many teaching activities and religious writings with my wife and lectures are noted. These remarks were written for personal remembrances with no thought of publication.

I was born in the small town of Ferriday, LA, near the Mississippi River, population of 2900. My father was the Baptist minister there for 20 years and my mother was a teacher who taught me 4 years of high school mathematics. When I was 12, to get me to read, my father gave me a Zane Grey western book he had in his library. I was hooked and have read Western books all my life. In high school, I played football, basketball, and baseball as well as singing in the church youth choir. Baseball was my first love. I practiced pitching until I could hit any spot on a brick wall where I had drawn a strike zone, wearing out numbers of baseballs. In my last year in American Legion Baseball, I won 3 and lost 1 as a pitcher and struck out 14 in one 7-inning game. I played baseball that summer with different teams from 9:00 am to 7:00 pm. In football, I practiced doing kickoffs for hours to become the kickoff specialist. In college I played football 4 years and baseball 2 years.

During World War II, I started my collection of army, navy, marine, and air force shoulder patches and now have over 1,000 patches. The rarest patch is for the Alamo Scouts, General MacArthur's private intelligence group. My wife gave it to me for our 50th wedding anniversary. In the 1990s, learning that part of the Battle of Nashville took place on the hill where we live next to Shy's Hill, I started collecting Civil War artifacts, some of which I found with a metal detector on our hill. I purchased or was given a number of Civil War pistols, rifles, canteens including a rare wooden canteen, and many other things. I also obtained a Japanese sword and pistol which were brought back by a Nashville army officer in charge of getting every Admiral and General to sign surrender papers, and give up their sword and pistol to signify their surrender.

Ferriday is famous for Jerry Lee Lewis and Jimmy Swaggart. A few years ago I saw Jerry Lee after his concert and as we were talking he said looking at me, "How many years was I ahead of you in school?" I said, "Jerry Lee, you were in the 8th grade when I was a senior." He said, "Did I fail that many times?"

From my commitment to Jesus Christ as a youth on, my Christian faith has always been a key factor in my life. After high school, I went to Mississippi College, a Baptist school in central Mississippi where my father had gone.

The contributions of MC influenced every important area of my life, my Christian faith, my wife, and my life's work. My four years at MC deepened my Christian faith and understanding so that when I faced new and tough questions of faith, that commitment saw me through.

Beyond the deepening of my Christian faith, the most important treasure college gave to me was my wife Jannelle. The year I graduated from graduate school, the BSU director at MC invited me to go with his group to the Baptist World Youth Conference in Canada saying he would introduce me to the most wonderful girl on the trip, Jannelle Landrum. So, I am enormously indebted to MC for the person who has made life so very special for me. We have two children, Melissa Dashiff and Chris Hamilton, and three grandchildren. Melissa, Duncan, and their children and my wife and I are members of First Presbyterian Church.

One of the things I much appreciated about MC was that I was free to try out for anything. I played football and baseball, and sang in the Concert Choir, earning varsity letters in football and music, singing in 47 concerts. One holiday when I was home, I sang at a church social with Harold Jenkins playing the guitar for me to sing. You may know Harold as Conway Twitty. One Christmas at a special program for the Women's Club by Amy Grant and Vince Gill at our church, I got to go on stage with a few other men to sing Christmas carols next to Amy Grant.

Football honed my persistence and toughness. My wife says I take persistence from being a virtue to a vice. In my first year in college, the freshman team ran plays against the first team. The first time I lined up at center, the nose guard in front of me, who was on the all-conference team the year before, said, "Coach says we are to go half speed but you and I are going full speed" and we did every play. In my junior year when I was playing first string center, we went to New York to play Colgate. The scouting report said the best defensive man on the team was a 230-pound nose guard over center. I weighed 160 pounds, the smallest man on our line. Believe me, it was a long afternoon. Few, if any centers can say they blocked their own team's extra point when I did not see a linebacker who bowled me over. However, the last two games were great for me. We had a center trap play in which the center faked left and then trapped right on the guard coming through. We made about 250 yards and several touchdowns on that trap play the next game. In the final game against Ouachita, we ran that trap play on the opening play of the second half, and our fullback ran 90 yards for a touchdown; the longest run in MC history. I also did the kickoffs for many games using a technique someone at MC before I had developed. The ball would bounce end over end and suddenly bounce very high. In one game the final back had the ball jump over him and go into the end zone. When he tried to run it out, we tackled him on the two-yard line!

The college also gave me a sense of mission and purpose in life in choosing a career in physics as God's calling. God's direction in that calling became more evident in my first year in graduate school at Indiana University. My mother said later when I came home the first Christmas, she had never seen me so despondent. I went back in January to my first semester finals not feeling good about my course work. Physics at Indiana was tough. Only 3 of 16 in the class before me continued on to receive a PhD and only 3 of the 15 in my class received a PhD. A week before exams, I was reading in the Book of Zachariah, Chapter 4, V. 6, and I read these words, "Not by might, nor by power but by my spirit saith the Lord." It was like when an Old Testament prophet wrote, "and the Lord said to me," I heard "fear not" "by my spirit you can do it." That did not lessen my preparation for exams, but when I went into those exams, my brain was firing like a giant engine.

The spring I graduated, Vanderbilt was looking for a nuclear physicist to replace a very research-oriented professor who was leaving to be department head at Michigan State University. He had just built the only iron-free electron spectrometer in the US, which was perfect for my research. He also left several graduate students who were working with him. That spring I received an NSF post-doctoral fellowship to go to Sweden for research and Vanderbilt made me an offer and appointed me to the faculty with a leave of absence to go to Sweden. Normally it takes a new professor several years to get funding, build equipment to do research, and attract graduate students. I felt God opened doors for me to step into a well-equipped laboratory with graduate students and an NSF grant. At the same time my closest friend from graduate school, collaborator under the same research director, and fellow Baptist, went to work at Oak Ridge National Laboratory. Through him, the nuclear laboratories at Oak Ridge were opened to me. I know of no other such research opportunities for a new assistant professor before or since. In addition, my Department Chair provided me with funds for the next two years to bring a more senior scientist from Sweden each year. In Sweden, I learned to do electron-gamma angular correlations. In my third year, I brought the Swedish scientist with whom I had worked and we built a magnetic lens spectrometer for such studies at Vanderbilt.

People who know me know that when I finished high school there were three things I would have liked to be a baseball pitcher, a cowboy, or a tenor. Physics was not even a word I knew. Also, people who do know me know I have been a lifelong Brooklyn, now LA, Dodgers baseball fan. Well, my wife gave me a 40th wedding anniversary present of a week at Vero Beach at the Dodgers fantasy baseball camp to meet and play with former Brooklyn Dodger greats Duke Snider, Ralph Brinca, and Carl

Erskine, and more recent LA Dodgers, Maury Wills, Dave Lopes, Steve Yeager, Rick Monday, and others. There were 2 highlights of the week, both in the game where the campers played the former Dodgers greats. Each of the 8 teams got one inning against the former Dodgers and the team that scored the most runs won a special T-shirt. In our last at-bat, I got a hit to drive in the third run to give my team the T-shirt award. Before the game, the outfielders from each team were all catching fly balls. One coach said can you beat the record of 23 caught without a drop last year. We said yes. Well, most of the balls were high fly balls where the player would move maybe 10 feet to catch it. I came up again at number 23 and the ball was hit way over my head. Running back at the last second I jumped up and caught it. The others gave a loud cheer and one yelled out, "and he's 69 years old." It was a wonderful week.

In 2007, I went again to the Dodger's Fantasy Baseball Camp. I played on a great younger team. In our last game, which was for the championship of our division, I was playing 2nd base. We were ahead one run and they had a runner on 1st and 3rd with 2 outs in the last inning. The next batter hit a hard ground ball that was about to go into the outfield. I dove flat out on the ground to my right, backhanded the ball, got up on my knees, and threw out the runner going to 2nd, with an underhanded pitch. My teammates immediately shouted "potato head award" which was given out each day for the most outstanding play. I received that award signed by all the Dodgers along with a trophy for my team winning the championship.

A number of years ago, a physics friend got me an invitation to join him at his friend's large ranch in New Mexico for a three-day spring round up of 500 head of cattle and branding of about 200 calves. After riding over 10 hours the first day rounding up cattle, I called my wife that evening. She said I sounded like a kid let loose in a candy store. It was a wonderful three days.

I guess my days in the concert choir at MC singing the first tenor will have to be the fulfillment of my third career choice.

Now, let me return to my real career. My year in Sweden with visits to laboratories throughout Europe and Russia introduced me to the world-wide nature of scientific research that was just developing and its importance to the world. My research has long involved international cooperation.

Also while in Sweden I bought my first sports car at a conference in England, an Austin Healey, and I have owned a sports car ever since, currently a Prowler.

In 1959 while in Sweden, J. Hollander from Berkeley and I visited laboratories in Leningrad and Moscow, the first American physicists since 1938 to do so. This began lifelong friendships and collaborations with Russian scientists. Walking around St. Petersburg in the snow and seeing small children in fur coats, I asked about their family class and was icily told there were no classes in the Soviet Union. I remembered that as we drove through the guarded gates to a house on the hills overlooking the Moscow River with a grand piano in its very large living room, the home of then Russia's only Nobel Laureate. While walking with a physics friend talking about world affairs, he said, "I spent 1,000 days in Leningrad when the Germans shelled us every day. So when I say I do not want war, I know why." Another evening, we were told a group of physicists would come to see us at 9 pm. We ate dinner at 7:00 pm and noticed a banquet table set up across the room. At 9 pm, we got up and went out to the foyer and were then taken to dinner by our Russian guests at the long banquet table where I effectively had two more full dinners. My colleague said I should have gotten the Medal of Honor for keeping up 7 US/Russian international relations. One of the physicists also said that night, "We do not want war. Why do you want war with us?" I said "We do not want war with you." Then he said, "But you have bomber and missile bases surrounding Russia, throughout Europe and Asia." Unfortunately, there is no connection between the desires of the people who did not want war and the government which tells the people what the government wants them to know.

In 1959, I attended the Baptist church in Moscow, visited earlier by Eleanor Roosevelt and Adlai Stevenson who signed the guest book. The service was two hours long and people stood in the aisles for three hours. After the service, I met with the leaders of the church and learned that Christians had played such an important role in supporting Russian troops in the Battle of Stalingrad that Stalin reversed his decision to ban churches in Russia and allowed them to open again. I also visited the Russian Mausoleum with both Lenin and Stalin inside. I was taken to the head of the mile-long line. People looked much longer at Stalin than at Lenin.

In January 1971, I was the first university physicist invited to address the Soviet Academy of Sciences. At the conference, Prof. Sliv, whom I met in 1959, came to my hotel at 10:00 pm to talk. So, that we would not be overheard, we walked the streets of Moscow until 1:00 am. I almost froze to death. On several occasions, I attended ballets at the Bolshoi Theater which were unbelievable and visited the Kremlin and many other places of historical interest.

In 1972, Jannelle and I went to a conference in the Crimea on the Black Sea. We were in our room only a few minutes when workers came to "fix the lights" but in reality to install a listening device. We had fun saying lots of crazy stuff. On that trip in Moscow, our Russian friends took us to a special restaurant at 9:00 pm. When we got to it, there was a long line but our host explained that we were Americans and they all stepped aside and let us go to the head of the line.

In 1973, I had a letter from a Chinese scientist, Professor Yang Fujia, in Shanghai asking for a copy of a paper I had written. I sent it and suggested we should get together to talk about our mutual interests in research. He said he would like that, but the time was not right yet. That was the middle of the 10-year cultural revolution in China in which many university scientists spent 10 years as farmers and factory workers. After the Cultural Revolution ended, he invited me to China in 1978 and I invited him and a delegation of Chinese Nuclear Scientists to visit Vanderbilt that fall. We entertained this first delegation of nuclear scientists in our home and took them backstage at the Grand Ole Opry. In 1978, I was the first American physicist to visit the Lanzhou Modern Physics Institute, their largest nuclear lab, which is over 2,000 miles from Beijing. They had me lecturing 8 hours a day for 3-1/2 days. These were recorded and published in Chinese and used by researchers and graduate students for the next 20 years. What was remarkable about these lectures was that never once did I see any person in the audience close their eyes or nod off. This trip was amazing. In Lanzhou, I was the only person in the western hotel and the chef would come and ask what I wanted for lunch and dinner. To get to Lanzhou, I took a 25-hour train ride from Xian through the Chinese interior. People were really shocked to see me, a westerner, whenever we stopped. On another occasion, we were coming down from the top of a small mountain when two young boy scouts were coming up.

They saw me, screamed, turned around, and ran. A short distance later I see them with two friends peaking around the corner to see this foreigner. I also went by car and plane to see the end of the 3,000-mile-long Great Wall of China which few had seen in the Gobi Desert. I visited the famous Dunhuang Mogao Caves with their marvelous paintings and carvings including their Mona Lisa done by Monks around 900 A.D. and covered over until the 1900s. I climbed a 1,000-foot-high sand dune to see a famous lake on the other side. In Lanchou, I was given an apple that tasted exactly like a banana. In the desert on the way, we ate watermelons grown from seeds taken to China in 1938 by US Secretary of Agriculture Henry Wallace. We returned to Lanzhou by a long train ride through the interior. As we climbed a mountain, people would run out of the villages with baskets to collect the unburned coal soot the train gave off to use as fuel. I also spent a day visiting a Buddhist monastery and attended their impressive services. The whole trip was amazing.

I attended a State Dinner given by the Premier of China for the visiting Premier of Nepal. The gymnastics were unbelievable. I saw Mao Tse Tung in his glass coffin. The last night in China in 1978, a dinner was hosted by the head of the organization that arranged all political visits to China, such as Richard Nixon's visit. During the evening, he said, "I called Lanzhou and asked if you did them any good. They gave you high praise for the lectures." Then he said, "But you and I know that these lectures were not the most important reason for your being in China today, the most important reason is that people who get to know each other can learn to live together in peace." That comment made a deep and lasting impression that has been a major motivation in all my international research cooperations. As the only Westerner I saw on the whole trip, I received red-carpet treatment.

In 1981, I went again for three weeks and again a booklet of my lectures was published in Chinese and used for the next 20 years. These two lecture series helped transform the Modern Physics Institute. On that trip I also got to see the tomb in Xian with all the soldiers and horses. In 1986, my daughter and her friend went with me to China for 3 weeks. I wanted them to see the factory where they made their famous double-sided silk paintings about 50 miles from Shanghai which I had visited in 1981. Prof Yang arranged for us to go on Sunday afternoon with his family. We toured the factory with the director and saw all the workers at their tasks. When we returned, the guide assigned to me said you understand the factory is not open on Sunday. Prof Yang got them to open it and have all the workers there to show your daughter their work! We were amazed.

I have seen much of China from Lhasa and seen Tibetan Yak trains driven by Tibetan cowboys, carrying salt from snow-bound Nam (Tengri) Lake at 17,000 feet in the Himalayan mountains, to a three-day cruise to see the three gorges of the Yangtze River before the dam was built. Lhasa is a very interesting city. We were there on a special day so the Dalai Lama's temple in the center of the city was crowded with people. There were many different groups of people with quite different dress. To leave Lhasa for the Himalayas, I had to have my Shanghai guide and the head of Scientific affairs get us out of the city gates. On that cruise, we stopped to see the famous Han Man who was buried more than 2,000 years ago in a liquid so that today his skin is pliable and every joint in his body works. Jannelle was unhappy that I did not bring any of that liquid back with me. On my 65th birthday, I climbed to the very top of the Great Wall of China which is a magnificent work across the very tops of high mountains near Beijing.

In the 1970s, I developed a close friendship with Professor Walter Greiner, a distinguished theoretical physicist at the University of Frankfurt. He came to Vanderbilt to teach an accelerated theoretical physics course for graduate students using his textbooks for 15 years. Working together, he developed a theoretical model of our discovery of nuclear shape coexistence which can be found in his advanced textbook on *Theoretical Nuclear Physics*. He nominated me to receive a Humboldt Prize of $33,000 in 1979 from the President of Germany for that work.

Indeed, international scientific cooperations have played an important role in world affairs far beyond anything the public knows. Examples follow in which I was personally involved.

In 1989, when university students in China staged the Tiananmen Square protest leading to the killing and imprisonment of students, Chinese universities worried that they would go back into another cultural revolution. Many Western scientists had been invited to speak at a physics conference in Beijing that next year. I and the other invited Western scientists wrote strong letters of protest to the government about Tiananmen

Square with refusals to go to that conference or any others in China. The government quickly changed their policies. Two years later I went to a conference in China and asked about why life at universities had returned to normal operations so quickly after Tiananmen Square. A high-level Chinese colleague said everyone was extremely surprised and noted two major reasons: (1) Western scientists had boycotted Chinese science and science is essential for China and (2) there were so many children of high-ranking Chinese officials studying in the US and Europe, mainly in the sciences, that they feared if they cracked down hard their children would not return to China. I had one such student.

Other examples involved Russia. In 1959, I became friends with a leading Russian nuclear physicist, Leon Peker. A number of years later the Russian government said it would allow Russian Jews to immigrate. The Pekers decided to apply. Unfortunately, the next day both lost their important jobs and were not allowed to leave. Leon contacted me and asked me to call him every few weeks, knowing we would be tapped. He wanted the authorities to know that someone in the US was very concerned about them so I did. Sometime later a visiting US Senator managed to convince the government to allow them to leave. They quickly went to Amsterdam. Then I arranged to cover their expenses to come to the US to seek a position. Brookhaven National Lab offered him a position in line with his leading work in Russia. After a few years, DOE decided to transfer him to my research grant where the overhead rate was much less. We collaborated in research for many years.

With the break up of the old Soviet Union, a major concern was the possible export of nuclear weapons and technology, including the scientists working with nuclear weapons to politically dangerous countries. My group had a collaboration with the largest nuclear laboratory in Russia and with a US Department of Energy laboratory in Idaho. The US government instituted a program to provide money to Russian nuclear scientists to keep them in Russia. Our collaboration got several hundred thousand dollars to help pay the salaries and research travel of a number of Russian nuclear scientists. Then with my urging my German collaborators in Frankfurt paid for several of our Russian collaborators to come to Germany for 2–3 months where they were paid German salaries worth several years of Russian salaries. Thus, a number of nuclear scientists have stayed in Russia to work on basic nuclear research rather than building weapons in countries very unstable and unfriendly to the US. I brought one of the nuclear weapons scientists to Vanderbilt to do basic nuclear research for two years.

During the Crotian-Serbian War, a physics friend from Zagreb wrote me that Serbia was threatening to bomb the old university nuclear reactor in Zagreb. He asked if I could get then Senator Gore to intervene to stop it. I contacted Senator Gore and he passed a Senate Resolution saying if Serbia bombed Zagreb, the US would swiftly retaliate. So, they did not bomb Zagreb and my friend sent me a letter of thanks.

In 1969, I started working to develop the first program in the US to study radioactive nuclei far from the stable ones found in nature and in 1970 initiated the University Isotope Separator Consortium in Oak Ridge, UNISOR, a unique consortium of 11 universities, and a state and federal government. It was not easy with the very tight university and federal budgets in 1970 to convince universities from outside Tennessee to put money into equipment and an operating budget at Oak Ridge for a university group as well as to convince the Department of Energy to match our funds. Our request to the DOE arrived at the same time that a major committee recommended the Oak Ridge Cyclotron be closed. Chairman Glenn Seaborg decided that since the universities were willing to put up money for a new facility and a five-year operating budget, DOE should keep the cyclotron open and provide the balance of funds for the next five years. We built the first magnetic isotope separator connected to a heavy ion cyclotron in the world. Some of our best research was done at this facility. This program also had a major impact on US nuclear science by keeping the Oak Ridge Cyclotron open so it could later be used to lower the cost as a part of a new national heavy ion laboratory.

This new heavy ion facility was to be different from all other nuclear facilities at DOE laboratories. It was to be a nationally used facility not just used by the local laboratory scientists. While serving on a national users steering committee for the new Oak Ridge facility, I saw the need for a new university institute adjacent to the laboratory involving Vanderbilt, the University of Tennessee, and the Oak Ridge National Laboratory to house university scientists next to the lab rather than 10 miles away in the city. While the Vanderbilt Provost did not approve my request for $100,000 to be matched by UT and ORNL for a university building, he suggested we meet with the Chancellor. In our meeting, Chancellor Heard said yes we do not normally have such funds but the opportunity for national leadership is so exciting that he would personally find the funds if the others came through. They did and we built the first building with sleeping rooms, a kitchen, and a lounge to form the Joint Institute for Heavy Ion Research. Dr. Freeman of DOE at the dedication of the new laboratory said "In nuclear physics, we have two outstanding success stories, UNISOR and the Joint Institute for Heavy Ion Research". In 1984, when I proposed adding a second building with offices, labs, and a conference room to make this an important conference center, Chancellor Heard hosted then Gov. Lamar Alexander for a breakfast where he got him to pledge $350,000 for the construction of a second building with additional funds from UT, VU, and UNISOR. With the approval of Chancellor Reese of UT Knoxville, I wrote Gov. Alexander asking for operating funds from the state to make the Joint Institute a Center of Excellence in Higher Education for Tennessee the likes of which the state had never seen. He did not respond but the next January he announced his new $20M Centers of Excellence in Higher Education Program which included funds for our Institute. Chancellor Reese said, "Joe, it is remarkable how much his new program matched your letter." Then Senator Henry got the Governor to help fund a third wing of the Institute. Our Institute has hosted over 100 international conferences attended by over 5,000 scientists and supported over 1,000 scientists from around the world to do research at Oak Ridge with Vanderbilt, U. Tennessee, and ORNL scientists. In 2008, at a conference on the 50th anniversary of my teaching and research at Vanderbilt, the then Deputy Director of ORNL Jim Roberto, gave a talk entitled *Joe Hamilton's Contributions to Nuclear Physics at ORNL*. The Deputy Director credited our Joint Institute saying, "JIHIR has played a pivotal role in the transformation of ORNL" citing it as the "First major user facility at ORNL, First joint institute and First example of a state-funded building at a national laboratory". When ORNL wanted to extend its mission, it proposed building three new Institutes: one for Biological, one for Neutron, and one for Computational Science. Our Institute in 1981 had already broken the barrier for a state building on DOE land shooting down all their arguments against this, until they said "Oh build that Hamilton

Hilton". Equally as important, when UT asked the state for $16M for the three new institutes, state officials were happy to do so because they knew of the great success of our Institute. These three institutes have brought in over $200 million/year in the transformation.

When Oak Ridge was competing for a new $1.3 billion spallation neutron source for condensed matter and biological research, U. Tennessee and ORNL used the success of our Joint Institute to get the State of TN to fund a Joint Institute for Neutron Science that helped DOE to propose ORNL as the site for the new 1.3 billion dollar accelerator. However, the Clinton/Gore administration had not approved the DOE/Oak Ridge request for the new accelerator. I was asked by ORNL to see if I could, through my contacts with Vice President Gore, get them to approve it. My nephew, Bill Purcell, was running the Clinton/Gore campaign in Tennessee for their second term in office. I explained to Bill the importance of the new facility. A short time later Bill called me and said he had spoken with Gore and explained the great importance of the new accelerator. In less than two weeks, Gore was in Oak Ridge announcing the construction of that new billion-dollar accelerator. When I took a group of high school students to see it later, the ORNL leader told them I was responsible for it being built.

In 1990, we wanted to build a new recoil mass spectrometer for Oak Ridge. We designed a large new generation RMS that would cost $4M but Argonne Lab proposed to copy an old design for $1M. Having a lower cost, they were funded. I set out to raise $1.5M from private sources so DOE would fund the balance because we would be able to do research Argonne could never do. I asked the State of TN for $500,000. With Senator Douglas Henry's support, we finally had a meeting with Gov. McWerter, Senator Henry, Rep. Bragg, a UT representative, and the Governor's finance minister. We made our arguments and the Governor looked at his Finance director and said if I sign off on this are you going to object? He responded, "last time I looked, you were the Governor". The Governor then said, "let's do it." After raising the $1.5 M, DOE then provided the balance of funds. In total, the new facilities and research funds at ORNL, in which I played a major role in bringing there, were 6.5 billion dollars!!

In the 1960s, my research with my graduate students and collaborators discovered the first evidence for the predictions of the Bohr–Mottelson model that beta vibrations of an American-shaped football where the nucleus vibrates in and out along its long axis to change its mean square radius would give rise to electric monopole electron emission, E0 but not the gamma vibrations. Then our group was one of three groups to discover that the first order B-M predictions for the relative gamma ray intensities from the beta bands did not agree with the experiments. One way out was to allow large M1 transitions with the E2 transitions. At the major nuclear structure conference in Tokyo in 1967, we presented our first evidence that there were no M1 transitions present. Mottelson in his invited talk said this was the first major failure of their model. Our research was made possible by a large National Science Foundation grant to our department for new facilities.

The most important discovery of our group in the 1970s and 1980s was that of shape coexistence in nuclei in part with the UNISOR facility. In 1970, every nucleus was believed to have one fixed shape as described by two leading nuclear theorists in Scientific American. Meyer and Jensen won the Nobel Prize for explaining the properties of certain nuclei which had spherical shapes, and Bohr and Mottelson won the Nobel Prize for explaining the properties of certain other nuclei with deformed, football shapes. In 1974, we discovered that in some nuclei these two shapes could coexist with separate bands and energy levels with slightly different energies, a heretical idea. When I presented our work at an international conference in 1976 in Russia, the idea was so heretical that the discussion went on for 30 minutes after my talk! A Russian friend remarked to me afterward that if you judged which paper was the most important at a conference by the amount of discussion that followed, then my talk easily won. Prof. Iachello from Yale said to me in 1985, "Joe, when you presented your work on shape coexistence at the Dubna Conference in 1976, most people did not believe it and those who did thought it occurred only in a few isolated cases. Now we see it throughout the periodic table." At the 2018 Nuclear Structure conference at Michigan State, the summary speaker said to me, "You must feel really proud here seeing almost every talk at the conference spoke about nuclear shape coexistence." Indeed, our work had changed the paradigm that each nucleus had one fixed shape to a new paradigm where every nucleus could have different coexisting shapes. As an aside, around 1990, a leading Russian scientist and subsequently a leading German scientist told me they had nominated me for the Nobel Prize in Physics for my work on shape coexistence that changed the paradigm that each nucleus had one fixed shape.

Now, scientists want to probe unknown nuclei still farther from stability that are important for understanding their nuclear structure and the processes that made the heavy elements in our earth. Also, they want to make new super-heavy elements with atomic numbers 116 to 126. This requires facilities to accelerate neutron-rich radioactive nuclei. It is tremendously exciting to move from 1970 when we started the first program in the US, UNISOR, to now where the Department of Energy is building a new, $760 million rare isotope accelerator for my field of research probing nuclei still farther from stability. Japan has a billion-dollar such facility now in operation. Many such facilities are being built in Germany, China, and other countries.

In 1990, I formed a collaboration of scientists from Russian and Tsinghua University in Beijing to study the nuclear structure of previously unknown neutron-rich isotopes of different elements. We had the largest data set from the spontaneous fission of ^{252}Cf in the world. Our group together has published more than 300 papers to greatly expand our knowledge of neutron-rich nuclei including the discovery of many new phenomena. These included octupole deformation (first in an even-even nucleus), hyper deformation, first coexistence of octupole and normal prolate deformation, chiral (right- and left-handed) doublet bands, region of maximum triaxial deformation, a new hot fission mode that evaporated 8–11 neutrons only in Ba and Ce nuclei and not others.

In 2005, my nuclear structure collaborator Yuri Oganessian, who was also the head of the Russian Laboratory that had discovered the new elements 114 and 116 and had preliminary evidence for 113,115 and 118, came to see me. To discover the new element 117 required a target of relatively short-lived ^{249}Bk Yuri had not succeeded in obtaining ^{249}Bk. Knowing of all I had done for ORNL, he wanted me to join him to help get a target of ^{249}Bk which could be made in the Oak Ridge high-flux reactor. The ^{249}Bk was needed to see if we could discover the new element 117. We then met with ORNL scientists who told us the target was prohibitively expensive but said there was a way to piggyback making our target on a commercial proposal. So for $600,000, we could get a Bk target chemically separated from commercial ^{252}Cf but they had no orders for ^{252}Cf on which we could piggyback. For 3 years, I called Oak Ridge every 3 months, and in August 2008, they said they had an order in progress. At the September 2008 conference at Vanderbilt, I asked Yuri, the Head of the Russian group, described to Roberto, the ORNL Deputy Director, the importance of this experiment. Roberto was convinced and had me and one of my ORNL collaborators write him a proposal for $500,000 for this new project. I raised another $100,000 from another of my collaborators from Lawrence Livermore National Laboratory. Our target was produced, chemically separated, and in the summer of 2009 shipped to Russia. I went back and forth to Russia during the 4-month experiment. In the experiment, a very rare and expensive neutron-rich isotope, ^{48}Ca, was accelerated to high energy to fuse with a ^{249}Bk atom in the target to make an atom of the new element 117, the first new element in over 5 years. Russia is the only country where 0.7% natural abundance ^{48}Ca is enriched to beam quality to make this experiment possible. These few rare atoms of 117 were separated and their decays observed. To observe 6 atoms of 117, we bombarded over a trillion, trillion atoms of Bk with 10 trillion, trillion ($>10^{18}$) atoms of ^{48}Ca. Our discovery of the new element 117 was reported in Physical Review Letters being highlighted with a picture on the cover in April 2010 after being accepted only three days after submission (normally it takes weeks to months). Publication was followed by over 250 news reports in major news media worldwide. I made the first report of the discovery of 117 at a large international conference in Canada. We continued to do research to definitively establish the discovery of the new elements 113, 115, and 118. Our collaboration was given the honor of naming the new elements 115, 117, and 118 which we had discovered. I was allowed to name 117 for the State of Tennessee, tennessine, because of my crucial work on its discovery. Prof. Oganessian has called me "the father of 117". Our work confirmed the 60-year-old prediction of a new Island of Stability around neutron number 184 and atomic number $Z = 114 - 126$. This was made possible by using very neutron-rich ^{48}Ca and neutron-rich ^{249}Bk to reach $N = 177$. The German and Japanese efforts only reached $N = 165$ and were much shorter lived. Plans are now being made for experiments to produce the new elements 119 and 120 and to make more longer-lived isotopes of 118 for study.

International cooperations with German, Russian, Chinese, Swedish, Dutch, Romanian, Korean, French, Polish, Yugoslavian, Slovakian, Italian, Indian, Turkish, Zambia, Uruguay, and Argentine scientists have played major roles in the success of our research along with our American collaborators. These cooperations were recognized by awards from the President of Germany, the President of China for Scientific and Technological Cooperation with PRC, Slovak, and Russian Academies of Science, honorary professorships at Tsinghua University in Beijing and Fudan University in Shanghai, and honorary doctor's degrees from universities in Frankfurt, Bucharest, two in Russia and India as well as three in the US.

In 1992, the State of Tennessee formed a Dose (amount of released material) Reconstruction Project to determine possible adverse health impacts to people living near ORNL from offsite releases of toxic and radioactive materials. I spent 8 years on the committee which met every 2–3 months. I read every page of our 3,800-page report at least three times. This study was the most in-depth and comprehensive study of such effects ever made. We did an invaluable service to the Oak Ridge community, the State of Tennessee, DOE, and the nation on possible adverse health effects in the past through the present from the release of toxic and radioactive materials. We received technical reports from the contractors and held regular public meetings to share what we had learned and seek their input. I was the only nuclear physicist on the panel so I played a critical role in the radioactive releases. In one presentation, I caught two important errors in the ^{131}I release report on their slides as they were being presented. The chair of the report said to me, "I have never seen anyone who can analyze in depth a slide and so quickly point out problems." He said after his final presentation he had first made his presentation to his team with five of them assigned to be "Joe Hamilton" looking for possible errors. Our report is invaluable for future environmental restoration projects.

I want to emphasize the great importance and contributions of the 64 graduate students who did their PhD thesis research with me, 30 MS thesis students, and the over 120 postdoctoral fellows and research associates who worked with me at Vanderbilt. In addition, there were 10 PhD students who used our data to do their thesis work at Tsinghua. These collaborations have formed an extended family for me, with many continuing to do research throughout their careers with me. Moreover, they have gone on to carry out distinguished careers in their own rights. For example, one has been principal investigator on two multi-billion dollar grants from the federal government!

My many research associates and collaborators from around the world have made major contributions to the research which we have carried out together. Of special importance has been the nearly lifelong collaboration with A. V. Ramayya. He came to work with me as a research associate in 1964 right out of graduate school. In 1971 he was appointed Assistant Professor and in 1981 Full Professor of Physics. While each of us has carried out some independent research, we have collaborated on 403 journal articles and 293 conference proceedings papers. Ramayya was the leader in our group in setting up and operating many of the experimental facilities in our research. In the 1960s, he set up the first Ge-NaI angular correlation system with multi-channel analysis and helped set up the UNISOR data acquisition system in the 1970s. He later worked with Professor Cormier at the University of Rochester to use his recoil mass spectrometer (the first in the US) and built at Vanderbilt a multi-neutron detector to use as a gate signal in the 1980s. The latter led us to build a new generation RMS at Oak Ridge. We co-directed many of our graduate students' PhD and MS thesis research as well as independently directing others. I believe our research collaboration has been one of the most productive and longest lasting in nuclear physics.

He made major contributions to many of our most important discoveries. I nominated him for the Beams Award, given for outstanding research by the Southeastern Section of the American Physical Society, which he received. He organized and chaired an international physics conference for my 60th birthday and 6 years later, I did the same for him. He has received honorary doctoral degrees from universities in Romania, India and the US and a Humboldt Award to spend a year at the University of Frankfurt in Germany. He was involved with me in helping organize the first six very successful international conferences on *Fission and Properties of Neutron Rich Nuclei* and co-edited with me five proceedings. For several years, he was the director of our laboratories for our introductory physics courses. He wrote successful manuals for these labs and used the income to endow an award for the best graduate student laboratory teacher. Ramayya and I and our families have been close personal friends throughout our lives. Other long time cooperations in nuclear structure research involved Y. X. Luo former Director of the Lanzhou Modern Physics Institute in China, J. Rasmussen UC Berkeley, S. J. Zhu Tsinghua U., Beijing, J. K. Hwang Korea, J. B. Gupta U. Delhi, India and Yuri Oganessian, G.M. Ter-Akopian Joint Institute for Nuclear Research, Dubna, Russia. Other long-term collaborations and friendships that have been very important were with Professors Lee Riedinger, U. Tennessee, Ed Zganjar Louisiana State U. and Russell Robinson, ORNL. I have formed lifelong bonds with each of these individuals.

I have organized and chaired 14 international physics conferences bringing together world leaders in nuclear research in different areas, beginning at Vanderbilt in 1964 and including six on Fission and Properties of Neutron Rich Nuclei on Sanibel Island. Michigan State University agreed to take on continuing this conference with me as co-chair of the seventh one which had to be postponed because of COVID and then a hurricane. Los Alamos National Laboratory has also been a co-host of 2016 and 7th conferences being planned.

I would like to give you a little flavor of some of my non-scientific trips. In 1982, I spoke at a conference in Spain and afterward went to visit a missionary friend in Kenya. One night we spent in the Tree Tops Hotel, where a young lady in 1951 woke up as the Queen of England. We saw a white rhino at the Salt Lick under the cabins and other animals. Afterward, we drove to Amboseli National Park where we saw thousands of zebras, giraffes, and other animals. We saw a pride of lions eating and one walked by our car where I could have reached out and touched his head. One went to sleep near our grass-topped hut where we were to spend the night. Looking for elephants, we found a giant bull elephant with tusks 6 feet long walking straight towards the car out of the bush. My friend drove out of his way so he would not smash the car. Then a young bull elephant came up and challenged the old bull to a fight for a few minutes before turning and running. Soon a second young bull came out of the bush and the first young bull got into a head-knocking fight with the second young bull. The old bull came up between them, wrapped his trunk around their two trunks, and ripped them apart chasing them off. I have a wonderful picture of the old bull looking straight at me doing this. My son went with me to a conference in Nambia where they are still mining diamonds and saw and heard Braying Jackass penguins.

For over 30 years I have given lecture demonstrations to students from elementary school to college to interest them in science. I started giving such lectures at Oak Hill Elementary School when my children were there. National Geographic did a story on my work with elementary school students for Geographic World for Kids. They came to Oak Hill to take their feature photo in the gym with my son's 3rd grade class for the article. Melissa, Chris, and their friends were in other pictures. Both also did their sixth-grade science projects with me, which we published. I ran a summer science program for high school students and teachers for 12 years doing different experiments at Vanderbilt and ORNL to help them see in action the different opportunities there were in physics. I also did two fifteen-minute physics programs for public TV that were shown nationally for 15 years multiple times a year. Friends from Oregon to Boston called to tell me how their children had enjoyed them.

In the course of my career, I have been fortunate to live and work in several different foreign countries, one year in Sweden, a half year in the Netherlands, over one year and two summers in Germany, and six months each in Russia and China on multiple trips. I have given invited talks in over 40 countries and visited over 100. The most wonderful parts of many of my travels are that my wife, children, and most recently grandchildren have joined me in these travels. In 1962–1963, Jannelle and I drove around twelve countries in Europe and lived in Amsterdam for six months. She went with me to a conference in Japan in 1967 following which we took a tour of Japan. When our children got older, we all went to a conference on a ship that cruised the Greek Isles. We spent time in Athens seeing the wonderful Greek antiquities. After the conference, we flew to Egypt where we visited the Pyramids and the Sphinx, rode camels, and visited the unbelievable antiquities in museums in Cairo and came back to visit Rome.

On our 50th wedding anniversary, we took kids and grandkids to our favorite destination, the Ocean Club on Paradise Island, Nassau. We stayed in the residence of the former owner of the island where on an earlier trip Melissa met the actor Sidney Poitier. Now, the Atlantis property at the other end of the Island has fantastic water slides, river rides, and other activities. On the way to the airport at the end, our grandson asked his mother, "Is there any word greater than awesome?" It was a fantastic time.

We offered to take each of our grandchildren to any place in the world they wanted to go when they graduated. Stephen, the oldest, picked Amsterdam, Paris, Normandy, and San Malo. We had a marvelous time visiting the Rijksmuseum in Amsterdam to see the works of Rembrandt including his famous Night Watch, visited his home and that of Anne Frank, and went to see the island villages in the Zuider Sea On to Normandy we visited the beaches, the High Cliffs the Rangers scaled on D Day and the cemetery. It was a moving experience. Finally, in Paris, we visited the Louvre Museum to see the Mona Lisa and other famous works and climbed the Eiffel Tower. Earlier my son had said to me, "Dad, you took me all over the world when I was young and it was very special. I hope you can take my daughter with you some." In December 2019, I took his daughter to a conference at the base of Mt Fuji. An earlier trip with the two children included 8 days in Tahiti visiting four islands. Jannelle twice went with me to Russia, once to Crimea and Moscow and once to St. Petersburg when I received an Honorary Doctorate degree from St. Petersburg State University founded by Peter the Great. There we saw the beautiful Palace of the Czars and the Hermitage Palace — now a museum. It is unbelievable, one room had 26 Rembrandt paintings in a row. These trips with my family have been the most wonderful and special parts of my life.

Finally, it has been a great pleasure to take my children to see and even meet their most favorite person. For Chris, it was to see Michael Jordan play in Chicago and in Nashville where he got to visit with him after the game in his dressing room. For Melissa, it was to go to the dress rehearsal for the Country Music Awards TV show one year and another time to go backstage to meet Randy Travis. For Jannelle, it was to have dinner in a small restaurant in Nashville with Robert Redford sitting two tables away. Over the years I met many of the stars of the Grand Ole Opry backstage.

I have taught more than 10,000 Vanderbilt students physics, including 26 of the 28 baseball players on their 2014 NCAA Championship Team

(I witnessed the win and personally congratulated all of my former students) and every member of the 2012 men's basketball team that beat Kentucky for the SEC championship. I have a baseball and basketball signed by each member. I also worked with the NFL Titans kickoff specialist who gave me a signed football. I taught many Vanderbilt student athletes, both men and women, in all sports including ones who won football Super Bowls, NBA Championships, MLB World Series with the Dodgers, and the $12 million FedX cup in Golf. One season I gave six lectures at six home games on the Physics of Football to 65,000 at the Tennessee Titans football games shown on the Jumbotron and seen by thousands more on TV.

In the late 1960s, Jannelle and I started auditing courses in the Divinity School at Vanderbilt. I decided to seek a master's degree in theology and I could take one course a semester. I continued to where taking one more course and writing a thesis would have completed my master's degree in Theology. This was important in helping us in our writing and speaking.

Throughout my career, I have been involved in many different religious activities. I taught a Sunday school class for 13 years for Vanderbilt and Belmont senior and graduate students and lectured at pastor's conferences, colleges and universities. In the 1970–80s, Jannelle and I wrote many articles on issues in science and the Christian faith including a booklet to give college students a firm place to stand both scientifically and religiously. Most of the articles sold over 300,000 copies so the total copies were over one million. A lady who teaches biology in N.C. wrote me a few years ago to ask if I still stood by what was written in the 1971 series of 7 articles for Baptist adults. I said I did and she replied, "In the 1980s when she was a biology major she was struggling with issues in science and faith. Her mother gave her these articles which her mother had kept for 15 years and they were important in helping her reconcile her science and faith." The biology teacher then said, "Now my son is in college and is struggling with whether to believe science or his faith." She wanted to give him these articles which she had kept for over another 20 years but she needed to be sure that I still agreed with them.

In 1971, I went on a 23-day speaking engagement to speak at churches and universities in Japan, Korea, Hong Kong, and the Philippines for the

Southern Baptist Foreign Mission Board. In Hong Kong, with no advance notice, I was told I was to be the featured speaker at a special dinner where they had invited 2,000 Chinese business and professional people at a downtown hotel to reach out to them for the Christian faith. The head of the Hong Kong Baptist Association said my presentation was the finest outreach they had ever had. I also spoke at the Hong Kong physics society and University and visited Macau. In Korea, I was speaking at a Baptist Church where men sat on one side and women on the other. While I was being introduced in Korean, I noticed the people nodding to something that was said. Afterward, I asked what that was that caused them to nod. He said, "I told them this man looks like a young kid but his father is 83 years old." The reverence for my father's age made it acceptable for them to listen to me. When speaking at Yonsei University, I talked about pear-shape nuclei. After that, I was told the Korean did not understand because the Korean pear is round.

Another activity in which Jannelle and I have been involved is through the Christian organization Compassion Internal. In 1968, we joined the organization to provide one child in Asia with monthly support. We have every year since supported one and then two and in recent years three children in Asia and Africa. In addition to monthly support, we write to each other now four times a year to share our lives, exchange pictures, and send money for their birthdays and Christmas and once a year to their families. This last Christmas we got pictures showing one child bought 3 goats, one bought 3 sheep and one bought a mattress and big bags of food. One year when our child and his family were about to lose their home and Uncle's place of business. We bought their two-story building and gave it to them. In another activity, through my cousin who was a missionary in Zimbabwe, we gave money three different times to build three churches in Zimbabwe after I visited there to see their work.

In 2017, the Senate and House of Representatives of the State of Tennessee passed a Resolution honoring my long career in teaching and research culminating in the discovery of the new element 117 and naming it for the State of Tennessee. California is the only other US state to have an element named for it. So, Tennessee will be recognized in every beginning textbook in physics and chemistry and in the Periodic Table of the Elements worldwide forever. The Metropolitan Council of Nashville

Davidson County presented me with a similar resolution. US Congressman Jim Cooper put such a recognition in the Congressional Record.

In closing, let me quote from the last stanza of the poem *The Road Not Taken* by my favorite American poet Robert Frost, "I shall be telling this with a sigh, Somewhere ages and ages hence: Two roads diverged in a wood, and I, I took the one less traveled by, And that has made all the difference."

Sanibel Poem*

Walter Griener

*Sanibel, Sanibel,
what a beautiful island
rich in shells,
and beaches with fine sand.*

*This is one of Gods best creations
ideally suited for recreation
and also, as we can see
for conferences with high fee.*

*Joe and Jannelle
once felt particularly well,
discovered many years ago:
Sanibel is the place to go!*

*And every year again,
to escape storm and rain
they come down from Tennessee
and enjoy the beauty of land, sun and sea.*

---

*In 2002, Walter Greiner wrote a poem about Joe and the conference he would organize
on Sanibel Island every 5 years.

Having friends world-wide,
they enjoyed with pride,
to move forward to share
tranquility, beauty, wildlife and air.

This happened 10 years ago
Celebrating 60 years of Joe
And since every fifth years
We come together there.

Now in 2002
we have assembled too,
and Joe is seventy of age
we congratulate and state:

Continue Joe, for many years!
You look-still young
and may become
near hundred: we shall cheer!

It is within your genes
Inherited from your father.
And, if I read: that means,
we may meet here often after;

At least for 6 more times
we shall enjoy progress in our science,
as we did in the past.
Long might it last!

Your father still wrote papers
At 98 of years.
Sure, you will match him later.
After all, you are his peer.

Jannelle was always on your side,
She gives you strength, she is your guide.
Treasure her, as you always did,
So that both of you are here indeed.

When we celebrate the next occasion,
streaming in from many nations
during 30 years ahead,
enjoying in '32 the greatest of all fest.

It will be interesting to see
where nuclear physics will then be.
Will the driplines be explored?
And the superheavies still adored?

Is the vacuum then understood,
will antinuclei, strange clusters have set foot
and arouse general attention
with their structure, I just mention.

That's all important to understand
galaxies, stars and distant land
at the edge of our universe
can we ever end this verse?

The future is full of promise
and indeed, we have great wish.
It's like looking into dark sky
and your thoughts do fly by
far beyond where we cannot see
and we ask, can we there be?!

We are in the hands of the Lord
And ask him for us to afford
happiness, health and insight:
May our all future be quite bright!

Let me now thank you all,
organizers, secretaries, here in this hall.
Enjoy still further Sanibel
and have safe journey home; keep well!

Happy Birthday to Joe

Tonight's task is to give Joe a toast,
also tweak his ego and make him roast.

And who better to use this roasting tool
than one who worked for him in grad school.

I am a person who enjoyed that fate;
I got my degree in sixty eight.

Our student club is large and vast,
and even now still expanding fast.

We all thank Joe for his trouble and pain,
but we feel that we survived a hurricane.

A whirlwind that comes to you swirling around,
needing illustrations on his way out of town.

But seeing him personally always got my vote,
instead of interpreting his handwritten note.

His handwriting is really beyond description,
unless he is actually writing Egyptian.

Before we discuss physics and Joe's huge part,
Let me tell you something about Joe at the start.

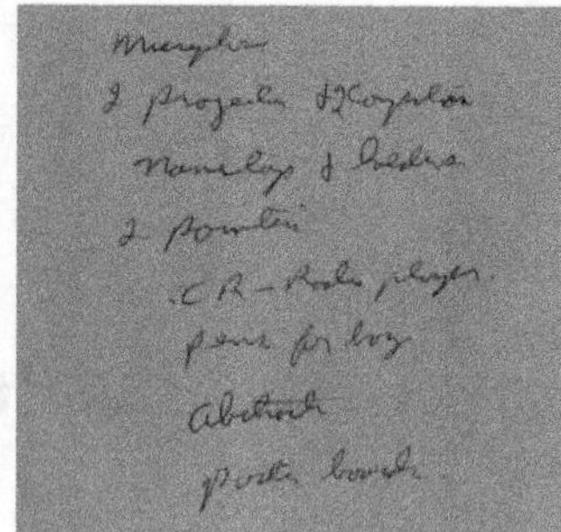
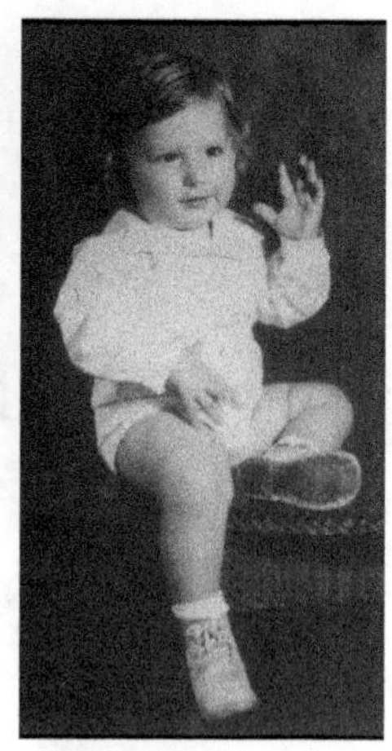

In '32 he was born in Ferriday Louisiana so wise,
In a town that produced famous and infamous guys.

Jerry Lee Lewis — a rock and roll Hall of Famer;
Jimmy Swaggart — a TV evangelist and known lady tamer.

A boy interested in things of all sorts;
Two big ones being religion and sports.

Long before science became his central stock
Joe was a baseball, basketball, and football jock.

He played all three sports which is a plan that's very
bold,
With the same energy we see now that he is eighty years
old.

Basketball — he was a member of the starting line,
A small but tenacious player wearing number nine.

Baseball — a pitcher and a little above par,
But its hard to believe Joe not being the star.

Football — short and tough was for him an easy
transition;
He was a starter at the important center position.

At Mississippi College he did it all it would seem,
Playing on the 1953 conference championship team.

There was advantage to being a two-sport college player
I assume it added to his image as a handsome lady slayer.

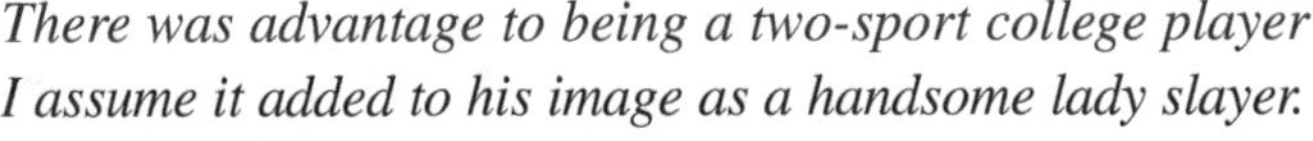

But sports and girls were not the only things to do;
Joe had time to study a little math and physics too.

For graduate school he went to Indiana U;
His interest in nuclear physics really grew.

As a postdoc he broke his strong regional bonds
And went to Sweden — the land of beautiful blonds.

But in Sweden he says he did not hook up with any local cutey;
Before he left he met Janelle Landrum — a Mississippi beauty.

Joe returned and joined the Vanderbilt faculty in fifty nine;
He thought staying a few years in Nashville would be just fine.

A few years has turned into fifty four,
And who knows just how many more.

They got married in nineteen sixty to start a magical match;
For Joe Hamilton, Janelle was clearly a major league catch.

His physics career took off like a shooting star,
But his interest in religion and sports did not stray far.

Joe's father was a life-long Baptist preacher,
So Joe became also a religion teacher.

For students, lay people, the public, and such
He has lectured about science with a religious touch.

But, luckily he did not require my thesis to take a stand
On the biblical significance of a beta vibrational band.

In baseball his hope before becoming an old codger
Was to play baseball as a Los Angeles Dodger.

He sometimes lays awake just before the sky turns blue
Wishing this magazine cover would just become true.

He has a Vanderbilt football jersey number one,
Which allows him to have dreams of sports fun.

When Vanderbilt is about to lose to UT as more of the same
Joe sneaks on the field, scores a touchdown, and wins the game.

In nuclear physics scoring a touchdown is not something new
As Joe has managed to save games and score quite a few.

Around nineteen seventy he had his first big score
By leading the formation of a project called UNISOR.

I wonder if you can spot in this photo a well-known pair:
A very young Carrol Bingham and Ed Zganjar with hair.

They well know the area in which Joe is extremely slow,
Is his understanding the meaning of the word no.

Saying "Joe that just cannot be done"
means that he has already half won.

When funds are tight and budgets flat,
Joe always figures out how to pass the hat.

Until you agree to help his project do well,
his energy might make your life a living hell.

100K from the state and 10K from each of you will suffice
And soon Joe has his UNISOR mass separator device.

Some people commit funds to keep his projects on track,
With the genuine hope that he won't soon come back.

And it works just like I have said
until the next project pops in his head.

Like a new accelerator, now please don't laugh
How about a twenty-five MV Van de Graaff!

Or, let's try an entirely new collaboration route;
A university — national laboratory joint institute.

Or, let's embarrass DOE in a major equipment stir,
Get state money for a recoil mass spectrometer.

Or, state budgets are tight so who would possibly figure
That Joe would succeed in making the Joint Institute bigger.

We used to wonder what would be next on Joe's grand strategy scheme,
Something really crazy — like discovering element one seventeen.

Just be aware if Joe comes to find you with his usual bounce
To ask you to contribute for Berkelium at two-eighty million per ounce.

If you think you can succeed by saying no
He won't leave your office, just will not go.

His boundless energy will not accept less
So in the end we all just tell Joe yes.

To extrapolate I can predict a coming strife
When St. Peter tells Joe its near the end of life.

Then Joe gets worked up into a considerable stew,
"I can't go yet — four more Sanibel events to do."

St. Peter realizes he has made Joe way too tense
And says "OK, I'll come back for you twenty years hence."

St. Pete doubts Joe will be publishing at age one hundred still
But most of us know Joe and feel he most surely will.

His students excel, retire, and go on the shelf,
But Joe keeps maintaining his Energizer Bunny self.

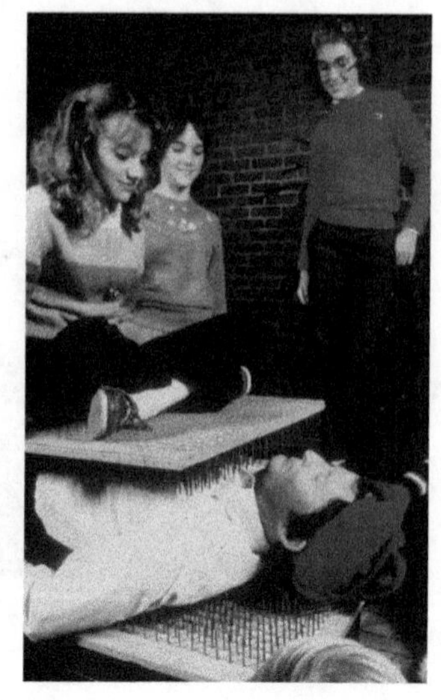

But, let us turn now to a very serious side;
There's a few things that we former students cannot abide.

Discussing the past — it's far too late;
We just ask you Joe to set some things straight.

Joe please try to be honest and fair;
You just have to be dying your hair.

And I don't want to cause a major rift,
But you've surely had at least one face lift.

It's those youthful looks of which you can be proud,
Compared to most of your grad student crowd.

Joe's his advisor, but it just cannot be;
Joe looks like a son of Bill Brantley.

And Joe's appearance is far finer,
With much more hair than Ed Zganjar.

The very worst thing is what everyone can see;
To the ladies Joe is far sexier than me.

I guess I have now given Joe enough grief;
It's time to express a personal belief.

Joe has had a vast grad student camp,
And on each he's left a unique stamp.

Now I have grown up and gotten old,
But find myself in the Hamilton mold.

I have long since fallen into one such trap;
Like him, my handwriting is not worth a crap.

I find myself far too much on the go;
Like him I don't often use the word no.

We all admire Joe for all he's done,
And thank him for his life on the run.

I am sure each of us can safely say,
That he's touched us all in a special way.

Most at age eighty long ago stepped aside,
But Joe continues to lead the whole tide.

We all have wanted to get Joe a gift of our
gratitude
Something that would fit his particular
aptitude.

We have thought long about what gift we could bring
For a person that already has every possible thing.

A plaque, shirt, or sweater, a trip to somewhere afar?
A hot rod, but he already has a Prowler sports car.

A suit, new book, a medallion for around his neck;
Then we decided to get him something new and high tech

I asked Ramayya, Carol Soren, and Walter Greiner;
All said that any other gift would be much finer.

When I asked those three to please explain why,
They all said that Joe is not a high tech guy.

But Ed Zganjar and I are not a couple of hicks;
We feel we can teach an old dog new tricks.

We decided to help you kick your anti-high-tech blues
And get you something even your grandkids can use.

I will stop the suspense and cut to the skinny;
The conference people present you with an iPad Mini.

We have loaded it with talks and photos from this week,
A conference momento to carry with you and keep.

We hope you like it and use it with vigor
To lighten your day of deadlines and rigor.

I only hope that you listen and heed my wise words;
Do not spend all day long playing the app Angry Birds.

If you do that your colleagues may think you foolish,
But your grandkids will really think you are coolish.

As a final thought before I end this verse and go,
For us all I wish you a very Happy Birthday, Joe.

Lee Riedinger
November 8, 2012
Sanibel Island, Florida

A Special Global Zoom Event
Science and Humanity:
The Extraordinary Life
of Joseph H. Hamilton*

Joe Hamilton retired from Vanderbilt University in August 2022, after 64 years of service. He has had an amazing life and career, as demonstrated in part by the following:

- Landon C. Garland Distinguished Professor of Physics Appointment;
- 10,000 non-science students taught;
- 1,200 publications;
- 100 graduate students with Vanderbilt degrees;
- 120 postdocs mentored;
- 450 invited talks;
- 13 international conferences organized;

---

*https://gradschool.fsu.edu/science-and-humanity-extraordinary-life-joseph-h-hamilton.

- 8 honorary doctorates bestowed;
- 21 articles on science and religion plus lectures to over 11,000 people.

His full CV can be **viewed here**.

This event was co-chaired by Da Hsuan Feng, Lee Riedinger, and Mark Riley. The event honored Joe Hamilton for an amazing life and allowed many of his close friends and colleagues to express their gratitude. This two-hour session included short presentations about Joe's accomplishments in four different aspects of his long career and was held on June 8, 2022, via Zoom from 9 am–11 am USA, Central Time.

The poster for the event can be **downloaded here (PDF)**.

The full video recording and other associated information can be found at https://gradschool.fsu.edu/science-and-humanity-extraordinary-life-joseph-h-hamilton.

Event Schedule for Main Video Feature (with times of appearance)

(Note that most presentations also have an associated pdf available by clicking on their respective names in the following).

Introduction: <u>Da Hsuan Feng</u>

Vanderbilt Dean of Arts and Sciences, John Geer (see separate video link)

Session 1: Science Endeavors (6 min 30 sec)
- Moderator: **<u>Stefan Frauendorf</u>**
- Speakers: **<u>Lee Riedinger</u>** and **<u>Ed Zganjar</u>**

Session 2: Teaching Endeavors (35 min)
- Moderator: **<u>Mark Riley</u>**
- Speakers: Bill Brantley and **<u>Sourish Dutta</u>**

Session 3: Global Outreach Endeavors (1 hr 0 min 45 sec)
- Moderator: **<u>Witek Nazarewicz</u>**
- Speakers: **<u>Alahari Navin</u>** and **<u>Yixiao Luo</u>**

Session 4: Fund-raising and Humanitarian Endeavors (1 hr 23 min 23 sec)
- Moderator: Lee Riedinger
- Speakers: **<u>Floyd Craig</u>** and **<u>Bill Purcell</u>**

Concluding Remarks: Joe Hamilton (1 hr 49 min 15 sec)

Encore: Open comments from attendees (1 hr 54 min)

Closing Remarks: Da Hsuan Feng (2 hr 22 min 20 sec)

Brief Curriculum Vitae: Joseph H. Hamilton

*(A full listing can be found at: https://gradschool.fsu.edu/sites/g/files/
upcbnu761/files/media/Files/Joe%20Hamilton-%20Mark%20Page/
HAMILTON_J_CV_Mar-2_2023_1.pdf)*

Personal: Born in Ferriday, Louisiana, August 14, 1932

History: Married Jannelle Landrum, Aug. 5, 1960; 2 children (Melissa C., Christopher L.)

Degrees: BS,1954, Mississippi College; MS, 1956, PhD, 1958, Indiana University

Dr. Sc. (Honorary) Mississippi College, 1982

Dr. Phil. Nat. Honoris Causa, Johann Wolfgang Goethe Universitat, Frankfurt, Germany, 1992

Dr. Phil. Nat. Honoris Causa — University of Bucharest, Romania, 1999

Dr. Phil. Nat. Honoris Causa — St. Petersburg State University, Russia, 2001

Dr. Phil. Nat. Honoris Causa — Joint Institute for Nuclear Research, Russia, 2004

Dr. Phil. Nat. Honoris Causa — Shukla University, Raipur, India, 2005

Dr. Sc. (Honorary) Berea College, 2007

Dr. Sc. (Honorary) Eastern Kentucky University, 2012

Professional Societies and Honors

Fellow, American Physical Society; Fellow, Am. Assoc. Advancement of Science; Am. Assoc. Physics Teachers

Jesse Beams Gold Medal for Outstanding Research, 1975, Southeastern Sect. American Physical Society

Alexander von Humboldt Prize, 1979 (Fed. Rep. Germany)

Harvie Branscomb Distinguished Professor, 1983–84 (Vanderbilt)

Featured in National Geographic World, Fun with Physics, 1983

George Pegram Award for Outstanding Teaching, 1988 Southeastern Section Am. Phys. Soc.

Earl Sutherland Prize for Excellence in Research, 1988 (Vanderbilt);

Winner of campus-wide student vote for Most Outstanding Professor at Vanderbilt, 1989; featured in Vanderbilt Yearbook

Guy & Rebecca Forman Award for Excellence in Teaching Physics, 1990 (Vanderbilt)

Professor of the Year in State of Tennessee by CASE (Council for Advancement & Support of Education), 1991

(One of the four finalists for National Professor of the Year, featured in their magazine)

Int. Symposium Nuclear Physics of Our Times, Sanibel Island, FL, 1992, Held in Honor of his 60th birthday

Thomas Jefferson Award for Outstanding Service to the University, 1995 (Vanderbilt)

American Association for the Advancement of Science 1996 Award for Development of International Cooperations

Jeffery Nordhaus Award for excellence in undergraduate teaching, 1996 (Vanderbilt)

One of 90 living Tennesseans with a biographical sketch in the Tennessee Encyclopedia of History and Culture, 1998 published in conjunction with the Tennessee Bicentennial Celebration

First Visiting Distinguished Laboratory Fellow at Oak Ridge National Laboratory, February 2000

First Outstanding Leadership Award, Oak Ridge Associated Universities, March 2000

Francis Slack Gold Medal for Outstanding Service to Physics, Southeastern Section, Am. Phys. Soc., 2000

Ilkovic Gold Medal for Career Achievement in Science, Slovak Academy of Sciences, 2001

Workshop Honoring his Distinguished Career, Univ. Georgia and Jefferson National Laboratory, 2002

Prize for International Scientific and Technological Cooperation with the People's Republic of China from PRC President, 2002

Flerov Prize for major achievements in Nuclear Physics Research, given by Russia, 2003

Distinguished Faculty Award, Dept. Student Athletics, Football 2009, Basketball 2012

International symposium held in Honor of 50 Years Teaching and Research at Vanderbilt, 2008

Elected Foreign Member of Academia Europea, The Academy of Arts and Science of Europe, 2011

Co-discoverer of the new elements 115, 117, 118, 2010–2012

Mentoring Award given by the American Physical Society, Division of Nuclear Physics, 2016

Resolution by Senate and House of State of Tenn. Honoring my long career and naming element 117 for the State of TN

Positions

Faculty, Vanderbilt University, 1958–present; since 1992 Landon C. Garland Distinguished Professor of Physics

National Science Foundation Postdoctoral Fellow, 1958–1959, University Uppsala, Sweden

Visiting Professor, Institute Nuclear Physics, Amsterdam, Netherlands, 1962–1963

Visiting Professor, University of Frankfurt/Main, Germany, 1979–80, Summer 1987, Summers 1990, 2001

Chair, Department of Physics & Astronomy, Vanderbilt University, June 1979–Sept. 1985

Adjunct Professor of Physics, Tsinghua Univ., Beijing, PRC, 1986

Honorary Advisory Professor, Fudan Univ., Shanghai, PRC, 1988

Guest Professor, Universite Louis Pasteur, Strasbourg, France, 1991

Founder/Chair, University Isotope Separator at Oak Ridge, UNISOR (consortium of 12 universities), 1971

University Radioactive Ion Beam, UNIRIB consortium, 1996–2015

Founder/Director, Joint Institute for Heavy Ion Research, Oak Ridge, TN, 1991

Director, NSF Summer Res. Participation for College Teachers, Vanderbilt, Summers, 1969–1971

Developed and directed two formal student exchange programs between Vanderbilt & Tsinghua & Fudan University, 1986–

Director, Vanderbilt Summer Science Collaborative Prog. for high school students & teachers, Summers 1991–2004

First Scholar-in-Residence, Hume-Fogg Academic High School, Nashville, 2006; Lecture for student body and classes

Organized and chaired 14 Int'l Research Conf: 1965, 1969, 1979, 1984, 1986, 1991, 1997, 1998, 1999, 2002, 2007, 2012, 2016, 2021

Over 75 Committees: National Academy of Science, American Phys. Soc., American Inst Physics, International Conf, Journal Editor

Member Oak Ridge Health Agreement Steering Comm, for the State of TN, 1993–2000; co-author of 3800-page report to State of TN.

Leader of major int'l collaboration with scientists in the People's Republic of China and Russia for 30 years and in Brazil, France, Germany, India, Poland, and Romania

Publications

Over 1,150 nuclear physics publications in refereed journals and books, with additional 6 published on teaching physics and 21 for general public on science issues. Co-author nine books: Science: Faith and Learning; Lecture Notes in Nuclear Physics (in Chinese); ORAU from the Beginning; How Things Work; Lecture Notes in Nuclear Physics (in Chinese); Graphical Representation of K-shell and Total Internal Conversion Coefficients from $Z = 30$–104; Modern Atomic and Nuclear Physics; Modern Atomic and Nuclear Physics, revised; Modern Atomic and Nuclear Physics — Problems and Solutions Manual; Co-author and editor 13 books: *Internal Conversion Processes; Radioactivity in Nuclear Spectroscopy; Reactions Between Complex Nuclei; Future Directions in Studies of Nuclear Far From Stability; Microscopic Models in Nuclear Structure Physics; Reflections and Directions in Low Energy Heavy Ion Physics; Structure of the Vacuum and Elementary Matter; Fission and Properties of Neutron Rich Nuclei; Perspectives in Nuclear Physics; Second International Conference Fission and Properties of Neutron Rich Nuclei; Third International Conference Fission and Properties of Neutron Rich Nuclei; Fourth International Conference Fission and Properties of Neutron Rich Nuclei; Fifth International Conference Fission and Properties of Neutron Rich Nuclei. Sixth Int. Conf. Fission and Properties of Neutron Rich Nuclei*

Lecture Series, International Conferences, and Seminars (in 49 countries)

Seven Endowed Lecture series on Science & Public Policy issues at Universities in US

Over 200 invited lectures at universities in US and abroad on public science issues

Over 450 Invited research talks at over 200 International Research Conferences in 30 countries

Six (7–24 days each) invited research lecture series at 14 universities in Asia and Europe (2 published booklets in Chinese)

Research Seminars at over 100 universities around the world

Senior Honors, Masters and PhD Theses Directed, Teaching and Outreach Programs

Directed one high school senior honors, 12 undergraduate senior honors, 30 MS and 64 PhD theses, and postdoctoral training of over 100 PhD's., and had 10 undergraduates in NSF Summer Research Experience for Undergraduates (REU)

Regular guest speaker for many years on noon radio talk show on science

Developed and gave lecture demonstrations for two 15-minute programs, one on electricity and one on states of matter shown 8–10 times a year locally and nationally on Public TV for over 15 years

For over 35 years given invited lecture demonstrations to local K-12 schools to interest students in science

Teaching emphasis is on physics for non-science majors. Taught over 10,000 undergraduate and hundreds of graduate students.

Modern Atomic and Nuclear Physics (Revised Edition), F. Yang and J. H. Hamilton (under-graduate-graduate-level textbook), World Scientific Publishing Company: Singapore (listed as their all-time best-selling textbook) (2010), 797 pages

In 14 years, the Vanderbilt Summer Science Program had 224 high school students and 112 teachers in a 3-week program doing experiments in a variety of different Vanderbilt and Oak Ridge laboratories, emphasizing careers in science

Testimony before Congress led to Department of Energy initiating summer programs for students at DOE national labs

The Joint Institute for Heavy Ion Research has brought 1,200, mainly young scientists, to work for extended periods with ORNL, VU, and UT scientists and sponsored conferences with over 5,000 attendees

The Joint Institute transformed Oak Ridge National Lab by opening the door for major State of Tennessee support for 3 new large institutes that bring in over \$200M/yr and thousands of scientists and students

Gave 6 different lectures on physics of football for 65,000 at 6 Tennessee Titans NFL games shown on Jumbotron during time outs

Gave 2-hour lecture-demonstrations to 30 high school students in the Vanderbilt Boys Exploring Science and Tech. Program for 8 years

A Tennessee Triumph:
Senate Resolution Recognizing
Dr. Joseph H. Hamilton

SENATE JOINT RESOLUTION NO. 1

By Senator Haile; Mr. Speaker McNally; Senators Dickerson, Yager, Harper, Yarbro, Bailey, Beavers, Bell, Bowling, Briggs, Crowe, Gardenhire, Green, Gresham, Harris, Hensley, Jackson, Johnson, Kelsey, Ketron, Kyle, Lundberg, Massey, Niceley, Norris, Overbey, Roberts, Southerland, Stevens, Tate, Tracy, Watson

and

Madam Speaker Harwell

A RESOLUTION
to commend Dr. Joseph H. Hamilton for his pivotal role in the discovery of element 117 and its naming as tennessine.

WHEREAS, it is fitting that the members of this General Assembly should pause to specially recognize those citizens whose laudable efforts have contributed to significant scientific breakthroughs; and

WHEREAS, one such person who evinces the greatest integrity in all his chosen endeavors is Dr. Joseph H. Hamilton, who played a pivotal role in the discovery of element 117 and its naming as tennessine; and

WHEREAS, a renowned professor of physics at Vanderbilt University, Dr. Hamilton officially proposed that the new element on the periodic table be named "tennessine" in recognition of the significant contributions made by Tennessee research centers Oak Ridge National Laboratories (ORNL), Vanderbilt University (VU), and the University of Tennessee-Knoxville (UT-K); and

WHEREAS, the impetus behind the discovery of element 117 grew out of a 2005 meeting at Vanderbilt University in which Dr. Yuri Oganessian, leader of the team from the Joint Institute for Nuclear Research (JINR), Dubna, Russia, asked Dr. Hamilton, with whom he had collaborated in nuclear research for fifteen years, to join him in obtaining a Berkelium (Bk) target to search for the new element with atomic number 117; upon visiting ORNL's High Flux Isotope Reactor, they learned that the most economical way to produce Bk was during the production of Californium (Cf); and

WHEREAS, in August 2008, when Dr. Hamilton learned the production of the Bk-Cf material was underway at ORNL, Dr. Hamilton arranged a collaboration with Dr. Oganessian and Dr. James Roberto, Deputy Director for Science and Technology, ORNL, to discover element 117; he subsequently invited scientists at Lawrence Livermore National Laboratory (LLNL), Livermore, California, to join the collaboration. If the venture proved successful, it was proposed by Dr. Hamilton that the new element be named in honor of the State of Tennessee; and

WHEREAS, working with Dr. Hamilton at Vanderbilt University was his colleague Dr. Akunuri V. Ramayya, who traveled with Dr. Hamilton to Russia to participate in experiments and collaborated in the analysis of data and in the writing of the scientific research findings; and

WHEREAS, superheavy element research conducted in Tennessee included the production and chemical separation of unique actinide target materials at ORNL's High Flux Isotope Reactor and Radiochemical Engineering Development Center, the production of which contributed to the discovery and confirmation of six superheavy elements and completion of the seventh row of the periodic table; and

WHEREAS, the discovery of tennessine, along with elements 113, 114, 115, 116 and 118, provides evidence for the long sought "island of stability," a concept that predicted nearly fifty years ago increased stabilities and much slower rates of decay for superheavy elements with higher numbers of neutrons and protons than those previously known; and

WHEREAS, Dr. Hamilton is a driving force behind Tennessee's emergence as a world leader for graduate education and research in nuclear physics; his fields of interest include experimental nuclear physics, nuclear structure, nuclei far from stability, heavy ion reactions, and superheavy elements; and

WHEREAS, born in Ferriday, Louisiana, Dr. Hamilton received a Bachelor of Science degree from Mississippi College in 1954 and a Master of Science degree and a Doctor of Philosophy degree from Indiana University in 1956 and 1958, respectively; he also received Doctor of Science (Honorary) degrees from Mississippi College in 1982, Berea College in 2007, and Eastern Kentucky University in 2012, Doctor Philosophiae Naturalis Honoris Causa degrees from Johann Wolfgang Goethe-Universität, Frankfurt, Germany, in 1992, University of Bucharest, Romania, in 1999, St. Petersburg University, Russia, in 2001, Joint Institute for Nuclear Research, Russia, 2004, and Shukla University, India, in 2005; and

WHEREAS, in addition to serving as the Landon C. Garland Distinguished Professor of Physics at Vanderbilt University from 1992 to present, Dr. Hamilton has also served as Director, Joint Institute for Heavy Ion Research, ORNL, from 1982 to present; Adjunct Professor of Physics, Tsinghua University, China, from 1986 to present; Honorary Advisory Professor, Fudan University, Shanghai, China, from 1988 to present; and Visiting Distinguished Laboratory Fellow, ORNL, from 2000 to present; and

WHEREAS, among Dr. Hamilton's many previous appointments are Postdoctoral Research Associate, Indiana University (1958); National Science Foundation Postdoctoral Fellow, University of Uppsala, Sweden (1958-1959); Assistant Professor of Physics, Vanderbilt University (1958-1962); a research appointment from the Institute for Nuclear Physics Research, Amsterdam, Netherlands (1962-1963); Associate Professor of Physics, Vanderbilt University (1962-1966); Professor of Physics, Vanderbilt University (1966-1981); Centennial Fellow, Vanderbilt University (1974-1975); Senior Alexander von Humboldt Fellow, University of Frankfurt/Main (1979-1980); and Chairman, Department of Physics & Astronomy, Vanderbilt University (1979-1985); and

WHEREAS, most notably, Dr. Hamilton initiated and established the first major research facility in the United States to study nuclei far from stability; creating unique partnerships with the State of Tennessee, the federal government, and public and private universities, he founded the University Isotope Separator at Oak Ridge, UNISOR, a consortium of twelve universities, in 1971; and

WHEREAS, in 1981, Dr. Hamilton founded the Joint Institute for Heavy Ion Research (JIHIR) in Oak Ridge supported by Vanderbilt University, the University of Tennessee-Knoxville, and ORNL; he is also the institute's first director, in which capacity he ably guided the institute from 1981-1987, and he has continued his astute leadership in that role since 1991; and

WHEREAS, in the 1990s, Dr. Hamilton helped design and raise the funds, including significant support from the State of Tennessee, for a new generation recoil mass separator at ORNL that opened up new areas of research there; and

WHEREAS, an active and dynamic member of his profession, Dr. Hamilton is a Fellow of the American Physical Society and a Fellow of the American Association for the Advancement of Science; he is also a Foreign Member of Academia Europes, the Academy of Arts and Science of Europe, a member of the American Association of Physics Teachers, Southeastern Section American Physical Society, as well as a member of Sigma Xi and Sigma Pi Sigma; and

WHEREAS, no stranger to accolades, this devoted professional is the recipient of numerous honors, including Outstanding Educator of America (1973), Jesse Beams Gold Medal for Outstanding Research - Southeastern Section of the American Physical Society (1975), Senior Alexander von Humboldt Prize - Federal Republic of Germany (1979), Harvie Branscomb Distinguished Professor - Vanderbilt University (1983-1984), Order of Golden Arrow Outstanding Alumni Award - Mississippi College (1985), George Pegram Award for Outstanding Teaching - Southeastern Section of the American Physical Society (1988), Earl Sutherland Prize for Excellence in Research - Vanderbilt University (1988), and Guy and Rebecca Forman Award for Excellence in Teaching of Physics - Vanderbilt University (1990), and was named the Council for Advancement and Support of Education's Outstanding Professor of the Year for the State of Tennessee (1991); and

WHEREAS, Dr. Hamilton has also been honored with the Thomas Jefferson Award for Outstanding Service to Vanderbilt University (1995), the American Association for the Advancement of Science Award for International Cooperation (1996), the Jeffery Nordhaus Award for Excellence in Undergraduate Teaching - Vanderbilt University (1996), the Francis Slack Gold Medal Award for Outstanding Service to Physics - Southeastern Section American Physical Society (2000), the First Outstanding Leadership Award, Oak Ridge Associated Universities (2000), the Ilkovic Gold Medal for Career Achievements in Science, Slovak Academy of Science (2001), and the Flerov Prize for Major Achievements in Nuclear Physics Research, Russia (2003); he holds the distinction of being one of ninety living Tennesseans honored with a biographical sketch in the Tennessee Encyclopedia of History and Culture (1998); and

WHEREAS, a selection of the over 1,100 publications of which Dr. Hamilton contributed to as an author or editor includes "University Isotope Separator at Oak Ridge: A Report of the UNISOR Consortium" (1974), "Evidence for Deformed Ground States in Light Kr Isotopes" (1981), "Effects of Reinforcing Shell Gaps in the Competition Between Spherical and Highly Deformed Shapes" (1984), "New Insights from Studies of Spontaneous Fission with Large Detector Arrays" (1995), "Modern Atomic and Nuclear Physics" (1996), "Octupole Correlations in Neutron-Rich 143,145Ba and a Type of Superdeformed Band in 145Ba" (1999), "Advances in the Spherical Shell Model: Shape Coexistence, Reinforcing Shell Gaps and Deformed Double Magic Nuclei" (1999), "Performance of the Recoil Mass Spectrometer and its Detector Systems at the Holifield Radioactive Ion Beam Facility" (2000), "Superdeformation to Maximum Triaxiality in A=100-112: Superdeformation, Chiral Bands and Wobbling Motion" (2010), and "Synthesis of a New Element with Atomic Number 117" (2010); and

WHEREAS, as a result of Dr. Hamilton's insight and initiative, an international team of scientists at JINR including VU, ORNL, and LLNL conducted the first experiment to successfully produce element 117; its discovery was confirmed through identical experiments conducted independently by GSI Helmholtz Center for Heavy Ion Research in Darmstadt, Germany; and

WHEREAS, having participated in every aspect of the discovery of element 117, on his next visit to Russia, Dr. Hamilton was introduced by Dr. Oganessian to his colleagues as "the father of 117"; and

WHEREAS, in December 2015, the Joint Working Party of the International Union of Pure and Applied Chemistry (IUPAC) and the International Union of Pure and Applied Physics (IUPAP) officially recognized the discovery of element 117, provisionally known as ununseptium, the second heaviest known element; as of 2016, twenty-one ununseptium atoms have been observed; and

WHEREAS, elements are permanently named for a deceased outstanding scientist or the person who made the discovery or for the region where the laboratories that made the discovery are located; to be so named is one of the highest honors in science; and

WHEREAS, in June 2016, IUPAC published a declaration stating that the Russian, ORNL, VU, and LLNL collaborators of the discovery had submitted their recommendation for naming the new element 117 "tennessine," with a symbol of Ts, after "the region of Tennessee"; and

WHEREAS, Tennessee is only the second American state to receive this distinction, one of the highest honors in science; Dr. Hamilton observed that "the name of the State of Tennessee will be in the periodic table in textbooks of physics and chemistry worldwide forever"; and

WHEREAS, "The new nuclei produced in this research have substantially increased lifetimes consistent with landing on the shores of the long theoretically predicted 'Island of Stability,'" said Dr. Hamilton. "These discoveries give evidence for the island's existence, and the new elements themselves represent a major advance in our understanding of the behavior of nuclear matter under the extreme stress of the ultra-large electrical forces that exist between the high numbers of protons that are packed into these new nuclei."; and

WHEREAS, throughout his estimable career, Dr. Hamilton has demonstrated the utmost professionalism, ability, and integrity, and it is fitting that the members of this General Assembly pause in their deliberations to applaud the remarkable achievements made by this exceptional scientist, public servant, and human being; now, therefore,

BE IT RESOLVED BY THE SENATE OF THE ONE HUNDRED TENTH GENERAL ASSEMBLY OF THE STATE OF TENNESSEE, THE HOUSE OF REPRESENTATIVES CONCURRING, that we hereby honor Dr. Joseph H. Hamilton for his prominent role in the discovery of element 117 and its naming as tennessine, commend his exemplary tenure as a professor of physics at Vanderbilt University, and extend to him our sincerest wishes for every continued success and happiness in his future endeavors.

Adopted: February 2, 2017

Senator Ferrell Haile

Senator Steven Dickerson

Senator Ken Yager

Randy McNally
Speaker of the Senate

Beth Harwell
Speaker of the House of Representatives

Bill Haslam
Governor

Congressional Record:
Honoring Dr. Joseph H. Hamilton

Congressional Record

United States of America

PROCEEDINGS AND DEBATES OF THE 111th CONGRESS, SECOND SESSION

WASHINGTON, FRIDAY, JULY 30, 2010

House of Representatives

HONORING JOSEPH H. HAMILTON

HON. JIM COOPER
OF TENNESSEE
IN THE HOUSE OF REPRESENTATIVES
Friday, July 30, 2010

Mr. COOPER. Madam Speaker, to a boy from Louisiana, the building blocks of life are food, faith, and a healthy dose of Southern hospitality. Joseph Hamilton, grew up to learn that our world is not so simple.

By January 2010 Hamilton had played a crucial role in an astounding discovery. He helped find something that no one knew existed. His work was critical in forming a multinational research team, and carrying out the discovery of Element 117—Ununseptium—the newest addition to the Periodic Table of Elements.

Hamilton's life started like that of any southern boy. Born in the humble town of Ferriday, Louisiana, he stayed true to his Baptist upbringing. He attended Christian Mississippi College, where God's calling led him to be a physicist. Hamilton then took his studies to Indiana University, where he studied nuclear physics and the elements. Their tiny atoms and their nuclei are invisible except to a select group of scientists with very advanced equipment. For everyone outside this elite group, the existence of atoms and their nuclei is purely a matter of faith. The only way to observe individual atoms of elements is through their impact on the world.

A skeptic may say that Christianity and physics, the two most important parts of Hamilton's life, cannot coexist, but Hamilton disagrees. He has pursued his passion without abandoning his beliefs, and has found that the two go gracefully hand in hand. As a professor at Vanderbilt University in Nashville, Tennessee, he and his wife have coauthored more than twenty papers on the harmony of physics and religion.

Professor Hamilton has dedicated himself to the growth of his students. Recognizing that they will soon take his place in research, Hamilton has supervised over 60 PhD dissertations and over 100 post-doctoral fellows at Vanderbilt. He includes his students in almost everything he does. One of his few regrets in his storied career is that he did not intimately involve his students in the discovery of Element 117.

Hamilton's research career at Vanderbilt over the last fifty-two years has taken him around the world. Russia, China, Sweden, and Germany have been but a few stops on his journeys. The creation of Element 117 in Dubna, Russia, just north of Moscow, was the result of a multinational project that Hamilton helped create. He believes that scientific discovery is a global effort, not a local one. Collaboration is key because science is one of the few things that unite us all in peaceful ways. Scientific principles apply around the world, regardless of race, creed, and nationality.

The first collaborative project that Hamilton initiated, the University Isotope Separator, has been a key operation at the Oak Ridge National Laboratory for more than forty years. This began as a collaboration of 11 Southeastern universities, ORNL and the State of Tennessee. Hamilton is also a founder of the Joint Institute for Heavy Ion Research, a cooperation of Vanderbilt, the University of Tennessee, and Oak Ridge National Laboratory. This Institute has become a world-class scientific resource. Moreover, this Institute opened doors that helped transform ORNL through the development of three major new joint institutes.

By January 2010, Hamilton's critical role in a joint Russian-American project came to fruition in the creation of six atoms of Element 117. While this new radioactive element has a half-life of only 78 milliseconds faster than the blink of an eye—its discovery points towards a fascinating possibility. Its half-life is longer than that of other recently discovered super heavy elements, and suggests that we may be on the path towards finding new, more stable, super heavy elements.

Hamilton and his coworkers' discovery will be forever emblazoned on the walls of chemistry and physics labs worldwide as the newest member of the Periodic Table of Elements. Generations of scientists will discover Element 117's properties, but no matter what is learned about Element 117, this Southern gentleman will always know that his work added to the building blocks of our world.

Metropolitan Council of Nashville: Honoring Dr. Joseph H. Hamilton

RESOLUTION NO. RS2020-540

A Resolution honoring Joseph H. Hamilton for his role in the discovery and naming of chemical element 117, Tennessine.

WHEREAS, the year 2020 marks the tenth anniversary of the announcement of the discovery of a new chemical element by an international collaboration of scientists including Joseph H. Hamilton of Nashville; and

WHEREAS, Joseph H. Hamilton, Landon C. Garland Distinguished Professor of Physics at Vanderbilt University, has had a renowned career as a nuclear physicist, and his role in the discovery of the new element was the culmination of many years of teaching and research in the field of nuclear physics; and

WHEREAS, Joseph H. Hamilton earned a bachelor of science in physics from Mississippi College and a master of science and Ph.D. from Indiana University, has been a member of the faculty of Vanderbilt University since 1958, and has taught and conducted research at other universities and laboratories around the world, with a focus on nuclear physics; and

WHEREAS, Hamilton has received numerous awards for his outstanding teaching and scientific research from educational, scientific, and international organizations including the Council for Advancement and Support of Education Outstanding Professor of the Year for the State of Tennessee in 1991; the Francis Slack Gold Medal Award for Outstanding Service to Physics, Southeastern Section American Physical Society, 2000; and the G. N. Flerov Prize for major achievements in nuclear physics research, given by the Joint Institute for Nuclear Research, Russia, 2003, among many others; and

WHEREAS, the new element 117, tennessine, was made by a fusion reaction of element 20, calcium, with element 97, berkelium, through an international collaboration including Hamilton' Vanderbilt Professor of Physics, Emeritus, A. V. Ramayya; and scientists from Oak Ridge National Laboratory in Tennessee, the University of Tennessee, Knoxville, the Joint Institute for Nuclear Research in Russia, and Lawrence Livermore National Laboratory in California. Hamilton was instrumental in the collaborative effort through his persistence in finding an opportunity for the production of berkelium at Oak Ridge National Laboratory. The discovery of the new element was first announced on April 5, 2010; and

WHEREAS, the discovery of tennessine has an importance beyond helping to complete the seventh row of the periodic table of elements. According to Hamilton, it provides evidence for the existence of the long-sought, theoretically predicted "island of stability" in superheavy elements that should possess increased stability and novel physical and chemical properties that could lead to new technologies, such as compact energy sources; and

WHEREAS, Hamilton initially proposed the name tennessine to honor the state where three of the research collaborators – Vanderbilt University, Oak Ridge National Laboratory, and the University of Tennessee – are located. The name was agreed upon by the entire team, submitted to the International Union of Pure and Applied Chemistry (IUPAC) in December 2015, and officially accepted by IUPAC in November 2016; and

WHEREAS, with the naming of tennessine, Tennessee became only the second U.S. state memorialized in the periodic table by having an element named after it, joining California which was the first, and the state of Tennessee owes Professor Hamilton a debt of gratitude for promoting this recognition.

NOW, THEREFORE, BE IT RESOLVED BY THE COUNCIL OF THE METROPOLITAN GOVERNMENT OF NASHVILLE AND DAVIDSON COUNTY:

Section 1. The Metropolitan Council hereby goes on record as honoring Joseph H. Hamilton for his many contributions to science and especially for his role in the discovery and naming of element 117, tennessine, thereby putting our state not just on the map, but on the periodic table of elements, where it will, according to Professor Hamilton, be seen in physics and chemistry textbooks worldwide forever.

Section 2. This Resolution shall take effect from and after its adoption, the welfare of The Metropolitan Government of Nashville and Davidson County requiring it.

SPONSORED BY:

Burkley Allen

Tom Cash

Russ Pulley

Brandon Taylor
Members of Council

A Glimpse into the Life of Joseph H. Hamilton

ORAU
OAK RIDGE ASSOCIATED UNIVERSITIES

Baldy
" GADZOOKS! THAT REMINDS ME OF ANOTHER EXPERIMENT!"

Is Prof.
Hamilton!!!

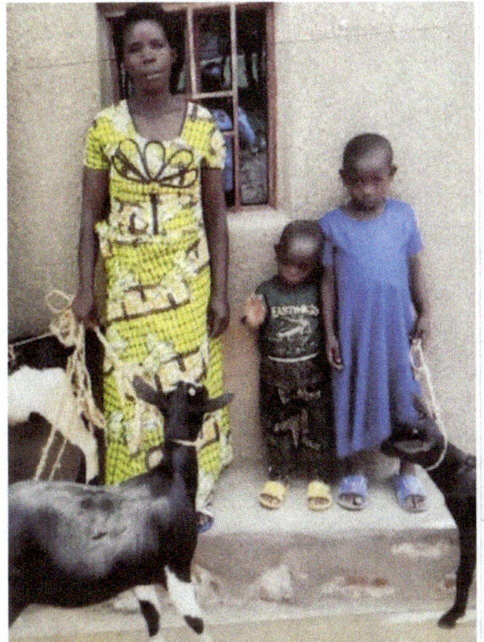

my mother was able to buy a 5 by 6
e were also able to buy food such
king oil, 5kg of sugar, 3kg of beans,
o able to buy a small gas cylinder.
e prosperity of good health and
the works of your hands.

Mississippi College
"Dixie Conference Champions"
1953
CHOCTAWS
NATIONAL CHAMPIONS
NATIONAL CHAMPIONS
Vanderbilt
VANDERBILT
MARTIN & ZERFOSS
INSURANCE AND BONDS
US Bank
Purity
Milk & Ice Cream
Golf at St. Andrieds

Index